JUSTICE AND THE EARTH

JUSTICE AND THE EARTH

Images for Our
Planetary Survival

———

ERIC T. FREYFOGLE

THE FREE PRESS
A Division of Macmillan, Inc.
NEW YORK

Maxwell Macmillan Canada
TORONTO

Maxwell Macmillan International
NEW YORK   OXFORD   SINGAPORE   SYDNEY

The Free Press
A Division of Macmillan, Inc.
866 Third Avenue, New York, N.Y. 10022

Maxwell Macmillan Canada, Inc.
1200 Eglinton Avenue East
Suite 200
Don Mills, Ontario M3C 3N1

Macmillan, Inc. is part of the Maxwell Communication Group of Companies.

Printed in the United States of America

printing number

1 2 3 4 5 6 7 8 9 10

Library of Congress Cataloging-in-Publication Data

Freyfogle, Eric T.

   Justice and the Earth : images for our planetary survival / Eric T. Freyfogle

      p. cm.

      ISBN 0-02-910695-8

      1. Environmental degradation. 2. Environmental policy. 3. Human ecology. 4. Sustainable development.

GE140.F74   1993

363.7–dc20   93–14389
                                                                           CIP

For Jane

The higher processes of art are all processes of simplification.

—Willa Cather

The Great Way is quite level,

but the people are much enamored of mountain ways.

—Lao Tzu

Contents

Acknowledgments

My work on this book was aided greatly by many people, principally the many writers over the centuries who have shared on paper their own thoughts about the issues that I address. They are so numerous that I could never list them all, and I am reluctant to take on the awkward, unfair task of singling out the most important few. Several people have been kind of enough to comment on parts or all of the manuscript, and to them I owe special thanks, particularly those who offered detailed comments while disagreeing with things I've said. J. Baird Callicott, Robert McKim, and Tom Ulen offered expert advice on particular chapters. Dan Farber, Bruce Hannon, Carol Kyle, and Don Zillman read drafts of the entire manuscript; their reactions helped clarify my thinking even if they did not lead me to avoid all errors. My editor at The Free Press, Beth Anderson, helped sharpen the book in many good ways. I should also like to thank H. Ross and Helen Workman, who through the Workman Research Grants of the University of Illinois College of Law provided generous support for my work and have helped make my academic home a far better place.

Perhaps my greatest indebtedness is to the many people over the years who by their examples have helped me see the land and gain a love for it. Here again I am reluctant to name names, either of those still living or of those now dead. They are best thanked, I sense, not by words but by deeds, by action that, in ways however small, helps continue their caring, never-ending work.

Introduction

My home county of Champaign, Illinois, near the center of the state, draws its name from the French word *campagne*, "countryside"—in this instance a broad, open countryside, one that reaches miles to the horizon in all directions. Visitors here are quick to draw generalizations about the landscape— that it is common, bland, and above all just plain flat. Yet a closer look at this land reveals that it is far from level, a truth that the first prairie settlers well knew as they surveyed the ground in early spring, imagining what it could produce and how they might tap it. Until the age of the power shovel and drainage tile, Champaign County was graced with expansive wet meadows and malarial marshes. When the rains came, nature's greatest force, gravity, made obvious the unevenness that adds subtle character to every local square mile.

Wherever we live, on plains or in mountains, on farms or in cities, the natural world around us is detailed and diverse, far beyond our capacity to absorb and interpret in detail. So great is this complexity that we couldn't begin to live with it if we didn't simplify, ignoring the richness of flora and fauna and leveling the land's dips and undulations. By simplifying, we create elementary images or conceptions of the world around us and of our place within it.

We carry these images with us every time we push open the screen door and step out into the world, whether to draw food, comfort, or excitement from the Earth's bounty. The most influential of these images are not of nonhuman nature alone, although these are important. They are images with people in them, images that depict us interacting with the land and its many living parts, images that help guide our actions and shape our senses of right and wrong.

This book is about how people interpret the land and their ties to it, about how they imagine their place on and within the colorful skein of Creation. How we live depends in large measure on how we think—not just individually but collectively, as a society and culture. It depends on the simplified patterns of understanding that we stow away within us, the patterns that we put on in the morning unthinkingly, like comfortable old clothes.

Humans make up the most intentional and thoughtful of all species, yet we haven't learned to live on Earth in a way that does justice to our planetary cotenants or to ourselves. Despite mounting evidence of planetary decline, we continue to behave as though we've been placed in a garden that we can feed upon and alter at will. We continue to act as if there will always be fresh pastures and new resources. We continue to rest our faith and to bet our futures on the cleverness of the human mind to solve the problems created by our overuse of nature's largesse.

My aim here is to gather together and talk about the mental conceptions and images that we now use, and ones that we might use, to guide our interactions with our fertile planetary home. Some of these conceptions are lodged in the subconscious and need to be flushed out and exposed to the light of day. Others are more conscious creations—by economists, philosophers, lawyers, and the like, who have used simple representations to explain how we now live and what needs to change. These images, too—some familiar, some less so—need a hard look, a look aimed not just at testing and criticizing but at finding useful, practical alternatives.

Our environmental predicament, I believe, stems less from our specific excesses and missteps in dealing with the Earth than it does from the misguided conceptions that we carry, from our collective lack of a guiding, energizing vision of sustainable life. Our principal problem, that is, lies in our thinking and imagining more than in our acting. What we need are new, more healthy images of our planetary role—images that inspire and are easily grasped, ones that can help us instinctively to separate healthy from destructive, right from wrong.

In talking about images of human life on Earth, I use the term *image* broadly rather than technically, in the popular sense of a mental representation, a conception, an idea. Some of our conceptions are easily grasped in visual terms. Others exist more in the form of ideas or simple narrative tales. What they all share is a radical simplicity: they all portray nature's complexity and the human role in elementary terms. My primary concern lies with the substance and wisdom of these images and conceptions; that being so, I spend little time looking into their history or talking about the cognitive and epistemological questions of how our senses gather data and how our brains interpret that data.

———

In the middle of the last century, Charles Darwin set out to explain how so many diverse species had come to live together on this Earth. Life on the planet seemed infinitely rich, with millions of species from sky to seabed, from ice floe to tropical forest. Somehow these species were all tied together, but how, he asked, and why?

Over the decades, step by step, Darwin's predecessors had made progress in coming to grips with this diversity. Darwin aided the quest by providing a few new bits of data for study. But he is remembered today not for his data but for his ideas, for his fresh way of seeing. The evidence was all there for Darwin's colleagues to sift through; they all held innumerable clues to the development of species through evolution and natural selection. What Darwin provided was a new frame of understanding, a way to piece the clues into a clearer picture of Creation and its many parts.

Darwin's story was new and startling, however familiar it sounds today. Species evolved, he claimed, bit by bit, into new, better-adapted organisms. They were related to one another through shared ancestors that lived eons ago, evolving through an endless process of competition and random change. Humans too were part of this process, cousin to the mouse and brother to the chimp. Humans arose in the same materialist manner as other species, struggling for food and safety in nature's endless, free-for-all game of life and death.

By offering a plausible mechanism for evolution (or, as he termed it, "descent with modifications"), Darwin helped put humans back into nature, in much the same way that Copernicus and his sun-centered astronomy had helped put the Earth back into the universe. But just as Copernicus left lengthy agendas for later astronomers, so too would Darwin leave his task incomplete. And so it remains today. In Darwin's simple vision—at least as we came to grasp it—the wolf and the zebra were merely distant relatives of ours, linked to us only in long-forgotten ways. To social theorists and to the general public, evolution meant inexorable progress. As the most evolved animals, humans by definition were the highest, the most exalted form of life—the natural rulers of land and sea.

In short, Darwin's evolutionary story, by offering only a distant, antiquarian link between humans and other species, succeeded only halfway in putting humans back into nature's order. People could admit their beastly origins and still hold to a comfortable belief that they stood above the rest of creation, somehow immune from the laws and vagaries of biology.

This book continues Darwin's work; not the scientific work of tracing evolutionary changes but the broader, probably never-ending work of conceptualizing the human role on Earth. Like Darwin's colleagues, we swim in clues. We see soil eroding, water becoming undrinkable, and species disappearing daily. Nature's resources decline in quality and quantity, and to those who watch closely the slide only seems to accelerate.

These clues, I cannot help but think, reveal a jarring discord in our relationship with the Earth. We've responded by plugging a few polluting pipes, preserving the last wilderness scraps, and so on. What we haven't done, at least not well, is to piece these clues together so that we can visualize our larger discontinuity and in some lasting, healthy fashion work to mend it.

Even with our massive and accumulated knowledge, we are still far from understanding our place on Earth. Scientists have learned a good deal—they have pieced together many of the clues. But even the best explanations leave crucial data unaccounted for. The clues left over, sad to say, are among the most important, for they are the clues that tell us the Earth is far richer and more complex than we realize. They are the clues that tell us that we don't know half of what we do.

———

My own thinking about nature began not far from where I now live, near the banks of the Sangamon River, which flows westward through the central regions of Illinois. The Sangamon meanders through the old land of the tall grass prairie and the oak–hickory forest. Except in winter, this lazy river runs brown, every year carrying toward the Mississippi a heavy load of rich organic soil, centuries of nature's work.

The river carries as well a heavy load of nitrates and other farm chemicals, especially in late spring. The native wolves, bears, elk, bison, and mountain lions are all gone today, and prairie flora once prevalent are now scarce or absent. Human–made reservoirs slowly fill with silt on their way to becoming mudflats, and groundwater supplies are threatened by toxic chemicals that percolate inexorably, invisibly, through the soil. With the plastic drainage tiles so effective, overflying waterfowl find the land less friendly, and the dwindling flocks now stop less often. Farms, homes, and factories are all addicted to heavy diets of nonreplaceable fossil energy, which means dependence on a distant lifeline that cannot and will not last. Genetic diversity dwindles yearly, here as elsewhere, as habitats give way to the pulsing, displacing human sphere.

These problems are unnerving, things to keep the thoughtful awake at night. Yet even they can seem distant and inconsequential on a cool September evening, when the ripe corn rustles in the breeze and the redwings and flocking grackles offer a choral accompaniment to the rust-streaked sunset. The cornfield yields a renewing aesthetic harvest, and we'd be remiss if we didn't put aside our troubles and reap it. Indeed, we are foolish whenever we dwell only on our environmental problems, on the negative and the harmful. If we are prone sometimes to ignore foolish resource practices, we're also prone to ignore the beauty, diversity, and resilience of our planet. The Earth deserves not just our restraint and respect but our wonder and applause.

—————

This book is offered to the reader who shares at least some of my confusion and dis-ease about our dealings with nature, to the reader who, like me, seeks more than just ways to recycle plastic and install low-flow shower heads.

It is also offered to those who suspect that the effort we once poured into civil rights issues needs to be duplicated in dealing with person-in-nature issues, which desperately need our attention. The kind of seething rage that motivated our 160-year civil rights drive is slowly building in this allied setting. In our interactions with the Earth, it is now the 1830s, and the prophets have few open followers. But more and more people are seeing the evidence and sensing that things are amiss; they're sensing that somehow, in some way, we need a broader definition of community. We need a greater

sense of justice in our dealings with the Earth; we need better images of a life that we can sustain for generations.

For many decades, American culture has exalted individual freedom as its highest value and goal. But freedom is not
an end goal, particularly when it means license to disturb
and destroy. Like equality, freedom is a tool to help achieve a
good, just life. As freedom is achieved, it becomes time to
start talking more loudly and deliberately about other components of the good life.

"State a moral case to a ploughman and a professor,"
Thomas Jefferson once asserted. "The former will decide it as
well, and often better than the latter, because he has not
been led astray by artificial rules." We can learn from acade-
mics and other specialists, but in the end we each must plot
a course, led by our hearts as well as our minds. We cannot
expect this important work to be done for us by scientists,
much less by corporations, however extensive their involvement will be. The problem is one of fundamental values and
of new forms of common sense, which means that we all
must get to work.

In the pages that follow, I set forth the results, so far, of
my own attempt to imagine a just way of living on Earth. My
account is necessarily a personal one, however quick and
liberal I've been in seizing and using the ideas of others. My
hope is that the reader will find here, not ultimate answers,
but intrigue, provocation, and at least pieces and glimpses of
a better way.

1

Sensing the Prairie

Unbidden, unplanted, the Old Prairie
Keeps humming the song of itself

—Dan Guillory

Nearly two centuries ago, the first white settlers pushed their way across the Ohio River into a place then known as the Northwest Territory. They sought open lands, where they could build their homes, sink their roots, and construct new lives. As they crossed the great wide river and spread into the new land, they gravitated toward the edges—toward the spots where the forests met the grasslands—just as the Native Americans had done before them. They moved as much by instinct as by forethought, for humans are edge creatures by nature, like the red fox and the white-tailed deer. Life at the edges mixed the best of both habitats. The forests offered fuel, shelter, and nuts; the prairies offered horizons, sunsets, and a sense of openness and grand scale. Best of all were the creeks lined with sugar maples—the oftnamed Sugar Creeks—which allowed them to make syrup to enrich the diet.

What the early settlers most needed was fertile soil for crops, and as they searched for good soil the prairie looked suspicious. Fertility went hand in hand with forest, at least as

1

they saw things. Land that produced only grasses simply had to be less fertile than land that gave rise to towering beeches and hickories. Land was valuable for what it could produce, and in the early appraisals the prairie got low marks. Not only was its fertility suspect, its tenacious root mass sternly resisted the horse-drawn wooden plow, to say nothing of the sweaty back and the hand-swung hoe.

Here and there—along roads, fields, and railways—bits of the native tall grass prairie still thrive today in the old Northwest. They've seen a lot, these aged relics, and with a little human forbearance they'll live on for centuries to come. A native prairie patch contains dozens if not hundreds of plant species, and its vegetative mass below the surface far exceeds the showy display it pushes eight, ten, and twelve feet skyward. By late August, a mature tall grass prairie supports ten million insects per acre. Plant life in the prairie is dense, and competition for space is so great that imported annuals like ragweed have little chance to take root. Even the nearby oak that drops acorns by the thousands can only hope that one acorn, some year, will get a start.

The native prairie remnants that stand today probably look much as they did to the pioneers, for the mature prairie changes slowly. Big bluestem, Indian grass, coneflowers, and switch grass all continue to press for dominance amid the colorful diversity of prairie life. Ground squirrels, mice, and ground-nesting birds still are populous and active inhabitants, their numbers fluctuating, as they always have, in response to forces that we little understand. Fire arrives here regularly—at least when people leave it alone—and insect hordes come and go.

If the physical prairie has changed little over the past two centuries, the prairie as an intellectual construct or vision has switched its clothing several times. The first settlers viewed it as a land of suspicious fertility, a place covered with dense, unfamiliar, seemingly useless plants. The open, treeless prairie bred a sense of isolation and despair among those who stood in its midst. By and large, it was an object of conquest, as if it had waited for millennia until the appointed day when humans would come to lay bare its roots. Its value, the measure of its worth, lay in its ability to satisfy human needs.

This new land was varied in ways far beyond the settlers' capacity to absorb, and they could respond only by transforming the richness of prairie life into crude, skeletal images. The hundreds of plant species quickly fell into a few categories, principally those that domesticated animals could eat and those that they could not. The settlers asked basic, elemental questions about the land: Could it be used for grazing? Could it be used for crops? Was it a safe and sensible place to live?

The first settlers dealt, not with the diverse, real-life prairie and its hundreds of species, but with the much simpler conception or image of it that they had created. Their images provided a shortcut, a way of carrying out daily life without succumbing to the prairie's overwhelming detail. Over time, the settlers began to revise their simple images. They learned by trial and error that prairie soil is far deeper and more fertile than the forest floor. Grass roots die and decay quickly in the prairie, and the rapid recycling of nutrients builds soil that is rich in minerals and organic compounds. As the settlers gained more knowledge of individual species, their simplified image gained more features. Some prairie plants produced illness in animals and were particularly disliked; a few offered medicinal or food value to humans and were worth seeking out. With the coming of stronger, nonsticking plows, the prairie root mass offered a less formidable challenge. Step by step, the prairie's stubborn image slowly gained a friendlier facade.

Imagining Nature

As we deal with nature on a daily basis, all of us are prone to simplify greatly, just like the settlers of the Old Northwest. We simplify because we must, because nature is so complex and our time to study it is so limited. Millions of species live on Earth—between five and thirty, the trained guessers suggest—and volumes could be written about each. The species constantly interact in greatly varied climatic conditions, and these interchanges could consume the pages of untold more volumes. One study of a native Nebraska prairie identified 237 plant species in one square mile. According to another study, the genetic information contained in the prairie would fill all the libraries of the world and then some. By necessity, we rely and act on the small bits of this knowledge that we glean from our reading, our listening, and our watching. Even this limited knowledge of nature is mostly stored away in the back of our brains, and we live from day to day guided by simple, easily used assumptions and understandings.

In the case of the prairie, humans have made use of several differing images over the years. When it represents good soil and farming opportunities, the native plants are plowed under *en masse*, and the soil community is exposed to the sunlight, perhaps for the first time in its history. When the prairie is viewed as land for development, plant and animal communities become irrelevant and are swept away. When the prairie is viewed as an object of beauty, it is preserved untouched, sternly adorned with warnings to "leave only footprints."

Most of us know few of the species that inhabit the prairie and little of how they interrelate. We are inclined to label them all simply as prairie grasses, including the legumes and other forbs. As we read, watch, and learn, the prairie gains more features—individual species, species with names, traits, and special beauties. We identify the mammals that call the prairie home, where they live and what they eat and fear. We begin to pick out the insects and birds, the mosses and fungi.

Our clarity of vision sharpens dramatically when we cut into the soil and uncover the richness that it contains, the earthworms that aerate it and break it down, the bacteria that fix nitrogen to promote new life. Prairie soil is a diverse life community unto itself.

We take another big step when we begin to study the species together. We see why some species thrive with certain neighbors and wither with others. We see that competition, cooperation, and symbiosis all are present. We become aware of the flow of energy through the system, from the sunlight hitting the leaf or stalk, to the moment the energy reaches the largest predator and is returned, through death and decay, to the living soil.

With time and education, we can begin to see the prairie in vastly different ways, and with this knowledge, the prairie is never again the same. It becomes much harder to treat it as part of the featureless mass of nature. Once we grow familiar with the mix of prairie species, we begin to spot differences among particular remnants, and we can rejoice at the discovery of a rare specimen. We begin to see how every day, in ways little and big, humans interact with the Earth's physical elements, with other species, and, indirectly, with human generations not yet born. Every act of plowing, building, and consuming reverberates into the natural world, well beyond the initial moment and initial place. Here a patch of prairie is loaded with nonnative annuals, which tell us of recent disruption. There a road crew has inadvertently mowed too wide and cut into a prairie remnant, and we may feel a sense of real loss, for we no longer perceive the plants as just a mass of weeds or grasses. Some plants can tolerate occasional mowing and may quickly regrow; others are more seriously harmed and will have trouble retaining their space. If the mowing is done in June, the prairie lover cannot help but think of the flowers that would have blossomed in July and August and of the seeds that, this year, will not mature.

Prairie preservation has come into vogue these days, at least among pockets of citizens where I live, which means that the prairie is gaining yet another face. The preservationist views the prairie as a work of art, as an object of intrinsic value to be treasured if not worshiped. Beauty is equated with freedom from human disruption, as if human-initiated change were inherently and inevitably ugly. In this new image, the prairie is a jewel of spare, subtle pleasures, evidence of unspoiled nature and of the possibility that humans can restrain themselves from altering every acre on Earth. The prairie holds and builds soil; it controls rain runoff; it provides home and food for hundreds of diverse species. The prairie is a busy, active place in this new image, rich and mature and with no need of the human touch.

If Kansan Wes Jackson has his way, we might all one day view the prairie and its native plants in yet another way—as a source of food. A botanist by training, Jackson and his coworkers at The Land Institute near Salina spend their days searching for perennial prairie plants that might reduce our reliance on corn, wheat, and other refined annuals. The crops now favored by farmers are planted annually, which means either regular plowing and erosion or no-till agriculture and massive chemical doses. Jackson is greatly pained by our nation's willingness to tolerate soil erosion, to ship several bushels of prime soil down the Mississippi for every bushel of grain we grow. His answer is to mimic nature by use of native perennials that avoid the need for yearly plowing, preferably perennials sown in mixed clusters to reduce exposure to diseases and insect pests. Jackson is one of the western pioneers of the ecological age, a modern-day Jim Bridger or Jed Smith—or more aptly, a modern John Wesley Powell—and his example will long shine even if the Illinois bundle-flower (one of his current favorites) never reaches our dinner plates.

Jackson and his colleagues may or may not find alternatives that woo us from our infatuation with high-yielding annuals like corn and soybeans. Yet, by searching, by questioning, they will undoubtedly be part of the process by which our visions of the land evolve. As we learn more from Jackson and others, the fresh-plowed field in spring will mean more than beauty and opportunity; it will mean exposure, erosion, and long-term soil loss. The endless expanse of wheat will mean more than harmony and wealth; it will mean monoculture, insect pests, and the inevitable pollution of farm chemicals. The center-pivot irrigation system, long a symbol of the blooming drylands, will mix its Eden message with a sense of draining aquifers and slowly poisoned soils. For Jackson, as perhaps one day for the rest of us, the prairie offers an image of sustainable success. The trick will be for us to homestead it properly.

Our Battered Image

Our culture's dominant conception of humans in nature has long been that of the human as conqueror and subduer of the wilds. Our ancestors were pioneers, driving out the wolves and turning up the soils and mineral riches. We are heirs of their myths and traditions.

This conqueror image is characterized by a high degree of teleological or ends-oriented reasoning. It presumes—we presume when we make use of it—that nature exists for the purpose of serving humans. Nature derives its value from its contribution to the utility of humans now alive, as if the field and the forest stood and waited for centuries for the human command to serve. We don't talk about this idea openly, of course, but it is plain from the things that we do.

With this image of conquest as our beacon, we feel free to distinguish between good and bad species, between the immediately useful and all others, and we take the knife, the torch, and poison to those that seem useless. The conquest image also simplifies nature in that it views each acre and resource as if it were discrete. We have long assumed, more or less implicitly, that our actions on one acre and with one resource will have no effect on the next acre. Guided by an image that blindly presumes this kind of unnatural separateness, we have trouble perceiving, and then acting upon, the overwhelming evidence of connections and interdependencies.

In our dominant cultural understanding, we humans are the subject, and nature is simply an object. This dualism is celebrated in our stories and songs, where we so rarely portray ourselves as temporary stewards and mere trustees. We sing aloud that "this land is your land, this land is my land" and that all of it "was made for you and me." Tellingly, our songs omit the fine print—about the obligations and dependencies that need to be part of our "rights" as "owners." "We all gather the pearls fast enough in this world," novelist Willa Cather once commented wistfully, "and nobody troubles himself much about the disease of the oyster which produced it."

This dominant understanding, with its many variants and applications, has suffered quite a beating as we've confronted more and more evidence of ecological decline. Our environmental difficulties have been hard to incorporate into our image of domination, for they are, it would seem, problems that shouldn't be happening. Nature is not supposed to tell people what to do. Its role is to provide challenges, excitements, and objects of conquest, but in the end its job is to submit. Nature is not supposed to impose nagging, intractable constraints on how we can live, on what we can do.

In facing environmental problems, our initial reaction has been to assume that nature is simply proving more stubborn an enemy than we first imagined. Our conquest has been incomplete, and through more aggressive manipulation we can complete the task of subjugation that began millennia ago. What we need, according to our dominant image, is more science, more technology—in short, more human cleverness. With enough cleverness, we can manipulate nature and do away with the apparent constraints that now confront us. With enough cleverness, we can continue to enjoy bluegrass lawns in the desert, grow corn on the hillside, and heat our leaky homes with fossil fuels. We can continue to consume resources, for technology will find more or will develop substitutes; we can continue to pollute, for technology will find ways to clean and contain.

This is how our culture has reacted, by and large, but we've been dimly aware in doing this that maybe we are asking and assuming too much. Our environmental problems are large, and those who watch closely see how they continue to get worse, despite busy scientists and lawmakers. Resource exhaustion is not a problem until shortages appear—except to those who watch the communal storeroom and are aware of the inexorable decline. Expensive efforts are made to control pollution, but the truth is that our control efforts largely involve trading one problem for another. We use scrubbers to clean our air but then end up with tons of slimy pollutants, which we then insert into the ground where they can move more slowly toward our waters and, perhaps ultimately, our bodies. Wastes once dumped in streams now also go into the ground, to create time bombs with a longer fuse. Nuclear reactors reduce air pollution, but at the cost of radiation leaks and with the legacy of high-level radioactive wastes. Even toxic dump clean-up efforts, which cost billions, very often involve shipping ruined soil from one location to another, where it may, in time, leak yet again.

We need to pull out our magnifying glasses and examine the Earth more closely. We need to pay closer attention to the omitted details. We need to reinsert our discrete acres and discrete resources into the interdependent natural world, where they belong. The little, invisible things are often our best clues—the oxygen levels in our lakes, the levels of phosphates and nitrates, and the presence of minute traces of pesticides. "Much of the damage inflicted on land," Aldo Leopold once wrote, "is quite invisible to laymen." It is too easy to ignore problems until they slap us in the face. Only a few environmental problems have slapped us so far, at least in earnest. Many more are gathering the strength to do so.

Once we start examining the details, we see more and more problems, and our cultural vision of humans controlling nature begins to transform into the more accurate, more dismaying reality of humans gradually soiling and spoiling the natural world. The process of learning to see can be a painful one. As Leopold confessed mournfully, "one of the penalties of an ecological education is that one lives alone in a world of wounds."

As we begin to rethink our dominant image—and the process is well under way—we face the problem that Darwin and his colleagues confronted. We have numerous clues, many of them just modest bits of data, and somehow we need to make sense out of them. Somehow we need to construct a core image around which we can assemble this mass of data. In some fashion we must develop a vision of the natural order that helps us learn how to live, day by day, without abusing our native surroundings.

When Darwin reexamined his data and developed his theories of evolution and natural selection, he helped restore the link between humans and other species. He showed that humans shared ancestors with other animals and hence had many cousins, albeit greatly removed. The step that Darwin took was important, but the link he established was dry and historical, if not obsolete. The Earth was inhabited millions of years ago, Darwin told us, by animals that would ultimately give rise to both humans and apes. Surprising news, shocking even, but the news only related, after all, to a state of affairs long past. Humans and apes might share distant ancestors, but the family lines had long since drifted apart. Like far-distant cousins we could easily claim that we no longer recognized one another. If our family ties with the apes were broken, who could take seriously our ties with more distant relatives like birds and fishes? Ironically and unintentionally, Darwin gave us a reason to be proud that we had risen up from the heap, that we no longer lived as our crude relatives did.

In the 1920s and 1930s, the new science of ecology added an important new strand to Darwin's historical tie. In scientific terms, ecologists explained the rather evident fact that humans are dependent on other species for energy and nutrition. Ecologists began to study the interactions of species, and they developed the notion that energy moves from species to species through complex food chains. Humans and other large predators stood at the apex of this food pyramid. To the ecologist, this lofty pinnacle meant dependence rather than superiority: because humans depend for life on all that lives directly beneath them, the removal of a rung could break the energy flow and knock down humans and others at the top. In the mid–1930s, ecologist A. G. Tansley coined the term "ecosystem" to describe the natural dependencies that bind humans to other life forms.

The ecologists added an important new element, but the only immediate lesson we learned from them was that we needed to protect our food sources. Cows and hogs, lettuce and oranges—the things we eat—these were the things that needed preservation. Wolves were pests, and moose and squirrels were edible but largely surplus and decorative. Insects and inedible plants we could poison at will, and the disappearance of a songbird or a wildflower appeared ecologically insignificant, however aesthetically painful. Ecology, in short, had little immediate effect on our long-held sense that some species are valuable and others are not and that we can distinguish between the two as we see fit.

In the popular consciousness, ecology meant little until 1962, when biologist Rachel Carson inflamed the public with her startling exposé, *Silent Spring*. Carson explained, coldly and logically, how a poison like DDT could move inexorably from species to species, gaining in concentration and lethality until it caused havoc. DDT scared the public, for it was a poison, Carson said, that never went away. Sprayed on plants, it passed by decay into the soil and percolated into streams and lakes. Small organisms absorbed the chemical, and it passed step by step into fish, which collected the poison in their bodies. When the fish were consumed by birds, the DDT stayed with the birds and altered their reproductive systems so that the eggs they laid had paper–thin shells. As the birds failed to reproduce, the once–musical spring turned silent.

In the years since *Silent Spring*, we've become familiar with the names of many strange chemicals. We are aware that mercury, dioxin, benzene, and many others can kill in small doses, whether reliably or sporadically, and that many toxins often linger for years. We fear radiation, global warming, and the breakdown of the stratospheric ozone layer. Deforestation and desertification now sound more scary, even when they occur hundreds or thousands of miles away.

Darwin and the ecologists took two vital steps. What now seems increasingly clear is the need for a third, more comprehensive step, a step that captures the ways that humans are inevitably tied to the natural world, a world we only partially understand. So long as we resist, as individuals and as a community, so long as we remain addicted to an image of nature as our supplicant, our little efforts to reuse and recycle will strike us as mere annoyances. And annoyances, like diets, are hard to take to heart. With a new vision, these same little steps can suddenly strike us anew; we will no longer be fleeing a past but striving for a future.

Building Anew

One image that appeals to many people is the preservation image of unaltered nature, of the native prairie patch untouched by the plow and the domesticated hog. The image has inherent appeal, for it offers a vision that is detailed and complete. It eliminates from the picture humans and their concerns. Once this erasing is done, all species can claim equal value and importance. As observers from the outside, we can relish nature's beauty without worrying about our needs. We can applaud the coyote because the groundhog it eats is not our dinner. We can marvel at the richness and efficiency of insect life when the plants being chewed were not sown by our hands.

When we study the prairie from this preservation perspective we see the rich varieties of life, and the prairie
regains its ties with the rest of the natural world. We see how
surface alteration produces erosion and lessens rainwater
control. We see how species are driven out by human acts
and how fertility and natural productivity decline with
repeated cropping. To the preservationist, the native prairie is
a place of beauty and harmony, a model of natural integrity
that shows us how species can live together on lasting terms.
Its processes are dynamic yet, in the long run, cyclical, sustainable, and health-producing—precisely the traits lacking
in the deteriorating landscapes where we now live. By offering a vision of sustainability, of nature's lasting health, the
preserved prairie offers a gauge with which to measure the
extent of human-induced change.

When we employ the native prairie as our guide for
proper land use, we become aware first of the wounds on
the altered prairie and then of the wounds that surround us
in all natural settings where humans have lived. These
wounds are painful and saddening, as if medieval master
paintings had been ripped with knives and splashed with
new paint. But the pain of existing destruction is by no
means the crux. Our concern soon shifts from existing
wounds to the continuing deterioration, which foreshadows
more wounds to come. The injuries are bad, but not as
alarming as the sense and evidence of gradual, inexorable
decline.

The appeal of the preservation ideal of nature is obvious,
and, as we shall explore in later chapters, its vision of
untouched nature can prove helpful in developing a new
image to guide our lives. But before proceeding, before gaining too much enthusiasm for the preservation image, we
need to see it for the incomplete image that it is.

Preservation is greatly flawed or, perhaps more accurately, has only limited value, because it omits humans from the
scene. The preservation image offers a vision of stability and
balance, but only by erasing people and, hence, by wishing

away the problems that we cannot help creating. As humans drop out, the prairie shifts from object to subject. This image might well prompt us to push for wilderness preserves, and preserves, to be sure, do yield irreplaceable value: wilderness is one thing we can never create. But preserving wilderness does little to make our cities more liveable; it doesn't bring our life practices into natural balance, whatever solace it might offer the guilty soul. The truth is, we can embrace preservation as our dominant land-use vision only if we think that our food comes from the grocery store and that we need not disturb the prairie to produce it. However much we love the wilds, the images of nature that we develop will work in daily life only if they have people and their aspirations in them. And these people must be living in a way that does justice to other life, to future generations, and to the physical Earth itself. That is the goal that lies ahead, and a picture with the humans wiped out won't get us there.

Before turning to some of the images that are being used today to portray our environmental plight, it may help to consider a few further points about the importance of this image-creating endeavor. Our environmental problems are hardly unknown to us. Indeed, we're bombarded with news about them daily. We also hear about efforts to deal with the problems—about clean-up efforts, stiffer penalties, new legislation, and the like. At times, in fact, the information flow seems endless, and a full-time job seems to await those who want to keep up. There are so many players and so much activity that it is easy to assume that our problems are being addressed, and we will soon be able to move on to other topics for diversion and news.

In several respects, the complexity of our environmental morass is right at the heart of the problem. And it makes the task of building good images all the more central. The world is complex; the issues we confront are numerous and interrelated. As an individual, I can quickly feel lost and helpless when I take the big step and decide to accept responsibility and do something. How is the average citizen to make sense out of confusing news about the ozone layer and global warming, much less figure out how to respond? Complexity breeds a sense of helplessness, which leads a person to push the problem onto others—big business and big government— and then to sit back, in anger and frustration, and fume. What can a person do in the face of such complexity except turn down the heat and recycle a box or a can?

Our helplessness is made worse because so much of our daily life is governed by others. The products we buy, the transport systems we use, the homes in which we live—all are made by others. We are offered choices, but they are limited. We have no choice when a fast-food restaurant offers coffee in a polystyrene cup, except to do without. In most neighborhoods, we'll have trouble building an earth-sheltered home, and we may even encounter resistance when we rip out our bluegrass and reseed our front yard with no-maintenance prairie plants. When choices are available, it often takes notebooks full of information to decide wisely. We can frequent a hamburger chain that uses domestic beef over one that fosters deforestation in Central America, but only if we keep close tabs on news stories about which corporation does what.

The point to be made about this understandable sense of confusion and helplessness is that our problems are societal in scope, which means that our solutions must be equally broad. Individuals today can adopt some practices that reduce environmental harms, but their range of action is limited absent a decision to drop out of society and return to the wilds. We can rise up to confront this complexity only by joining together, as a community and a people, to demand practices and products that are more sane, harmonious, and sustainable. We cannot all grab microscopes, nor can we realistically read the detailed reports of researchers and investigators. What we can do, what we must set out to do as our part of the solution, is to lay down broad guidance on what we see as the way of living rightly—the good life, if we might call it that. We can build a consensus on what we

value and what we want and then demand that our governmental and economic structures help us get there.

What we are talking about in part is a greater concern
over the ethics of right living, and we could talk about the
task (and will do so in later chapters) in moral and ethical
terms. But the typical reaction to talk of ethics and morals is
that it involves lists of "shall nots," things that are confining
and restricting rather than liberating and empowering. Most
of us are uncomfortable with preaching and moralizing; it
seems to tell us only what we cannot do. What would serve
our needs better is a vision of this task that provides us with
a challenge and a goal; images that create "a seedbed in the
heart," as Stephanie Mills observes, so that our new ecological practices can "root, branch, and flower." If we can do that,
if we can develop a vision of better life that we can reach for,
we are more likely to make the effort to do so.

A well-drawn image should stimulate and organize our
aspirations. It should provide a vision of justice and a benchmark to use in measuring right and wrong; it should serve as
a glittering grail or a new Jerusalem, as a spiritual context, as
a focus that gives meaning to our life, if we pursue it with
vigor. Love, awe, and reverence must all be part of it. "There
must be a mystique of the rain," offers theologian Thomas
Berry, "if we are ever to restore the purity of the rainfall." At
bottom, our image must provide a new measure of value.

As we undertake to imagine a new way of life, we must
realize that the goal of dealing justly with the Earth is not
like the goal of putting a human on the moon. It is not a
goal that we can establish, turn over to scientists, and forget
about, even though the contributions of scientists will no
doubt be immense. This goal, instead, will be more like the
goal of achieving racial justice, for the task will require social
adjustment more than technological inventiveness. What lies
ahead is a messy, time-consuming, wrenching process of
growth.

Our first changes will be behavioral, as we learn new
ways of dealing with the Earth's physical components and

other life forms. But gradually we will build a new outlook on life, a new sense of what is right and what is just, and it will permeate us, it needs to permeate us, down to the core. In the words of Aldo Leopold: "To change ideas about what land is for is to change ideas about what anything is for."

2

The Economics of Overuse

Man's power over Nature means the power of some men over
other men with Nature as the instrument.

—C. S. Lewis

Some time ago, I bought with my wife a little over one hundred acres of land in the countryside of east–central Illinois,
about forty miles south of the university town where I teach.
As farmland goes, our land is poor stuff. The soil is heavy
laden with clay, and only in good years will the best acres
produce corn like the rich humus soils further north. The
land is partly wooded, and the terrain is unusually broken
for this part of the state. The road that runs along the land is
a modest dirt affair, something that, as best I can tell, has
only lightly felt the roadgrader's steel blade. Even this dirt
road dwindles into obscurity as it passes our land and enters
a neighbor's field. On our land the sounds of human life are
faint and few.

A creek wanders through the property for a third of a
mile, and its banks contain a respectable water flow—on
occasion even a powerful flow, when the rain is heavy and
the upstream drainage lines send their waters down unnatu-
rally fast. Along either side of the creek, which is thickly

19

lined with willows and silver maples, lie some thirty acres of bottomland. The soil here is blacker and richer, but regular flooding compacts the dirt and deprives it of the loose, crumbly structure that fosters root growth. The creek is a demarcation line of sorts; not far from its south edge, the land rises up nearly forty feet to an undulating meadow. The hillside, gentle here, steep there, is dominated by mature trees—spreading white oaks, arching hickories, and a surprising number of flowering hawthorns. Some owner decades ago may have deliberately planted the many haws, but their origin is, and is likely to remain, one of my land's many mysteries.

On an early visit to this place, before I had bought it, I spotted a coyote den dug into the hillside, so big that I couldn't help but think wistfully of bear. The black bear, though, has been absent from Illinois since about 1860, at least as a regular inhabitant. Two similar dens cut into the thick bottomland by the creek. At most one of these coyote dens would be in active use at any time, given the territory needs of this thriving scavenger. I spotted the dens in the spring—birthing time for local coyotes—on a day when the white-flowered hawthorns and purple prairie phlox colored the awakening land. Not long thereafter, the resident appeared in person to check me out. I spotted the coyote across one of the bottom fields and for a few moments returned its motionless, focused gaze. This, I could see, would be one of my cotenants, and I was glad of it.

Many motives led me to buy this land. The acquisition of a source of income was not one of them, or at least not an influential one. Still, some of the acres were flat, and the seller, an energetic young grain farmer, made clear his desire to continue tilling these acres by shares. A third of the tilled land I decided to remove from cultivation to devote it full-time as food and home for wildlife. The remaining plowed lands, the forty acres best suited for grain, I agreed to lease out, as much to please my amicable seller as for any thought of gain.

The county court records tell me that I now stand as the latest in a lengthy chain of occupants of this special land, dating directly back to 1837 when one Thomas Clawson and one Reuben Dannals took government patents for it. About earlier native users of the land, the records are silent. Over the recorded decades, the owners have come and gone, and boundaries and land uses have shifted to and fro. Cabins and farm buildings have also risen and fallen, or at least so say the occasional government survey maps, which reflect only dimly the human drama that has taken place here. I am intrigued by the varied building sites marked on these old maps, for the land as I first saw it contained not so much as a run-down hog shed. One day, maps and spade in hand, I'll search for the plowed-over evidence of early foundations, as if to defy for a moment the larger forces of time. Even our homes and hearths, it seems, are soon mere dust.

Today I'm considered the newest steward. In time, though, I'll pass the land along to others and with it will go the rights and duties that accompany the office of landowner. One day I'll join Clawson and Dannals as one of the early owners, if centuries from now our names are remembered at all.

When I bought this land, I didn't know what to do with it. What I did know was that I wanted my reign of steward-ship to be one of waxing rather than waning land health. The need to think about particular uses confronted me before I had even struck a deal to buy the land. I needed to set a price for my offer; I needed, that is, to translate these 103 acres of field and woods, coyote and stream, into a par-ticular number of dollars to insert into a small blank on a lengthy form. I knew what I could afford, but this number of dollars said something about me and nothing about the land that I sought. I knew as well what the seller was asking, but his asking price was his number, reflecting his way of think-ing about the land and his way of sizing up its attractions and its opportunities.

I came up with a number, as the occasion demanded, but

I was uneasy about the idea and the process. The dollars I offered and paid had no magic to them; they contained none of the life and the beauty that the seller on the other side had to offer. If truth be known, I added nothing to my offer of dollars because of the resident coyote and the mystique that it brought. The creek with its beaver and muskrat—these too only reduced my offering price, for wooded floodplain is largely worthless for the grain producer whose needs so dominate the local land market.

In some economic sense, I suppose, there was an equivalency between what I paid and what I got. But how do we compare life with nonlife, the natural with the artificial? How much is true, and how much is untrue, in the equation that says that, on the day of our closing, my money equaled the seller's small corner of the Earth?

————

One of the main ways that we think about the land, and about our dealings with the land, is in terms of dollars and cents. This land is worth five thousand dollars per acre; that land is worth only one-tenth as much. This urban lot, nicely located, goes for fifty thousand dollars for a mere fifth of an acre. The timber on this hillside sells for one number; the mineral rights sell for another. The plentiful air is free to use; the neighborhood park, being publicly owned, also costs the user nothing.

Dollars and prices are, most obviously, the terrain of economists. Money is one of our near-constant preoccupations, and, inevitably, several of the most dominant images of our environmental plight come to us from economists and related observers. By and large, what we hear from them is that imperfections in the free market are at the core of our environmental problem.

Maybe our problem is that we aren't pricing things right. Maybe our problem, brought down to scale, is that we just don't know how to put a price tag on the coyote, the crooked creek, and the patches of purple phlox.

The images that economists have to offer are simple for the most part. Like most simple images, they rest on key assumptions about human nature and the capabilities of human knowledge. Given the importance of our image-building task, we need to grab these economic images by the scruff of the neck and give them a good shaking to see what they're worth as guides for daily living, whatever value they might have for technical discourse.

The Permanence of Land

We can start with the simplest economic–related conception of the land, one that comes initially and largely from the accounting profession. The received wisdom of accountants is that land is a permanent asset, an asset that retains its value and productivity and that therefore should not be written off gradually or depreciated over time. Buildings and equipment have limited useful lives and are exhausted, slowly or quickly, like the old cabins and sheds on my land. The land, however, stays in place and will always be with us. The county court records seem to bear this out. Clawson and Dannals took my land in 1837; someone not yet born will be in charge in the year 2137.

This image of land's permanence is preached as accounting gospel, with only modest limitations. It's a solid image, something to lean on. The farmer who plows his hillside and suffers erosion is reassured by this image, for in calculating profits and losses at year end, he is told flatly to think nothing of erosion losses. His grain storage bin may slowly wear out, and his expense calculation must show this deterioration, but the land lives on, he is told, as an ever–producing asset. This image of permanence also reigns in the federal income tax code. Here, too, the general rule is that land is an asset of unlimited useful life. In the gospel according to the IRS, land use generates no deductible costs.

The permanence of land is a gross simplification, of course, and we use this simplification and others because they are convenient. Our norms cannot incorporate every factual nuance without becoming unwieldy, so we pare away. The tax code treats land as permanent, and the net effect may be beneficial, given that fair revenue collection—not sound living—is the revenuers' sole goal. But for our other norms, those that we use as guides for conduct, this generalization has terrifically high costs. The land that I bought is not the same land that Clawson and Dannals first plowed during the reign of Martin Van Buren, and I can spot many of the probable changes.

Only in the most trivial sense can we attribute permanence to the land. A parcel takes up space on the Earth's surface, and we can always define its geographic coordinates. But this is not what we mean by land.

Land is the basis of life, the place that we inhabit and where we build our homes and schools and farms. The land joins with the seas to form the foundation of the biosphere, that thin, fragile, transient layer of the Earth's crust where we find all known life. Farms are worthless without topsoil and the millions of organisms contained in each gram of it. As a people we have become highly skilled in pushing our soil out to sea. My land has not escaped this process. Small gullies mark the most obvious exit routes that my soil has taken. The creek in the spring season, thick, churning, and brown, is the escape route for the rainfall's upstream thefts.

In many other ways, the land shows its vulnerability. The irrigation of desert lands concentrates salts and minerals on the surface. Over the years, the salts build up and the crops stop growing, which is why thousands of acres of farmland annually are taken out of production—more or less permanently. Salinity ended Mesopotamian irrigation centuries ago, just as it causes irrigators in California's Coachella Valley to use much of their water not to serve crops but to flush away salt left by earlier irrigation—a slowly losing effort. Our pollution and land-use practices give rise to desertification, faster in North America than in Saharan North Africa. Our overgrazing of cattle and sheep causes grave surface deterioration. With our high-tech chemical industry, we poison the soil and drain its fertility, and our toxic wastes render some land suitable only for artillery target practice. Nuclear radiation offers even more potent tools of destruction because it covers wide areas so quickly. The soil is the source of life, and, as the saying goes, we treat it like dirt.

The notion of land's permanence is a considerable and risky simplification. Yet, even so, it can be valuable if we turn it to new uses. The image of permanence suggests that the land will always be around and will always be a part of our lives, respect or abuse it as we might. It is the principal with which we work, and we must recognize and use it as such. As our permanent fund, land cannot be treated as a depleting asset, for if we let it disappear, it is irreplaceable. There is, then, an undeniably right implication to this image: no more land will be created; we must make do with what we have.

If we do not already treat the Earth as our principal, as something we must respect and not destroy, we must soon begin doing so. The truth is not that land *is* permanent, but that we must change our actions so that it *will be* permanent.

Those who regularly assume the land's permanence commonly limit this assumption in several ways. Accountants tell us that nonrenewable resources like coal, oil, and minerals are all exhausting assets, and in accounting for their use we must recognize their gradual depletion. When we begin to extract an exhausting mineral, we must estimate the extent of the deposit or reserve and then determine the exhaustion rate. Term by term, we must monitor the depletion and include as an expense the value of what we consume.

The discipline and rigor of this requirement is invigorating, for it slaps our cheeks with the reality that we are consuming something that one day will run out. It forces us to drop all pretense that we are using the income rather than the principal. A similar accounting rule applies to human alterations of the landscape, such as contour reshaping and tree planting. These changes as well have a limited useful life, and our accounting image requires us to recognize their limits.

When we think about how depletion accounting works, we might wish that the practice of depletion enjoyed wider usage and that it cut more deeply as an exception to the assumption of the land's permanence. Perhaps the farmer who loses soil to erosion and exhaustion should regularly confront the fact that his behavior differs little from the oil company that pumps a limited reservoir. Perhaps the irrigator whose soil builds salinity should face the same truth. When we recognize that we are depleting, we see more plainly what we are doing.

But, having seen, how will we respond? Will we be shocked at the truth and alter our behavior? Would my predecessors on my land have acted differently, would they have left unplowed my rolling meadow, if they had been required to keep records of the land's decline?

The risk in using depletion in this way is that the landowner can soon get used to the idea of depletion and accept it as normal. If we dutifully account for soil loss every year, perhaps we will soon accept the loss as a typical cost of doing business. Depletion better embraces the truth of what we are doing, but it might serve poorly as a prod for better behavior. Much the same ambiguity and danger lies in the larger idea of land's permanence. If the image serves as a goal and we reach out with it to make land permanent, we can put the image to good use. But we are as likely to use the image to drift into complacency: If the land is permanent, we don't need to protect it or even worry about it.

The Common Pool

A second economic depiction of our environmental problem arises out of what is known as the "common-pool problem." A common pool is an asset that is used concurrently by many people, each of whom has the right to increase his or her use without limit. The paradigm illustration of the common pool is the old town commons in England and early New England, where villagers grazed their animals. In the simplified version of this tale, each villager could add more animals to the common and thereby grab a larger share of the forage. An unregulated fishery offers a more contemporary and probably more accurate example of a common. The fishery is used by many, and each fisherman can grab a larger share of the catch by using more nets, lines, or traps.

The common-pool problem carries with it the story of the tragedy of the commons. The tragedy comes about when individual users of a common increase their use and thereby bring about a reduction in overall productivity. As more animals are grazed, the common grazing lot deteriorates in quality, and the amount of forage produced declines. As more fish are caught, the breeding stock is reduced and the annual fish production goes down. The tragedy comes about because individual users are motivated to increase their use, even if they anticipate the decline. The individual fisherman adds a few more nets because the extra fish taken are all caught and enjoyed by him, whereas the corresponding loss is shared by all the fishermen. The individual may end up ahead, even if the overall catch declines.

Many observers believe that this common-pool story with its tragic conclusion captures a good part of our environmental predicament. Their suggestion is that too many of our resources are held and used in common, which prompts users to abuse. Our air is a common asset, and factories pollute it, in part because the costs of polluted air are shared with all who must breathe it. Many of our groundwater aquifers are declining rapidly and will soon face exhaustion, again because use is communal and each pumper has no incentive to be the one to stop.

According to many economists, the solution to the tragedy of the common pool is to divide the common asset and distribute shares or parts to individual users. The grazing lot can be divided into discrete lots that grazers will own individually. The fishery can be allocated by setting an overall quota on the annual catch and then giving each user an individual limit. Once this "privatization" takes place, the incentive that leads to the tragedy is removed, or so the argument suggests. Now each grazer must accept the full costs of overgrazing; if he puts too many animals on the lot and productivity declines, the loss is entirely his own. The assumption is that the grazer will exercise foresight and discretion and will limit use to a sustainable level. The fish quota system operates even more simply, for each fisherman is prohibited from exceeding the sustainable catch.

This common-pool analysis, with its tragedy and implied solution, is readily transformed into a simple image of the natural world and the human role in it. In this image, nature is a collection of economic resources available for human use. We overuse some resources because they are shared. This tragic tale helps explain why air and water pollution are so prevalent, why aquifers and fish populations decline, and why we each are reluctant to cut back when local water supplies are short. Each common has its limits, including our national parks and wilderness areas, which are easily overused by recreational visitors. This overuse of communal resources is a cause of our environmental problem, perhaps even the most important single cause.

The implication is clear: when we are faced with a common pool, we somehow must each recognize the limits and ensure that we do not exceed our fair share. The implication is further that our environmental problems would largely end once privatization occurred and that no problem exists in the case of assets already in private hands. In drawing this conclusion, economists assume that people act to promote their economic well-being as they see it. Rational people, implicitly, do not abuse or destroy assets under their exclusive control, which is why privatization is the answer.

After offering its important insights, however, the common-pool story runs into troubles, particularly in its suggestion that our best course is to divide nature into privately owned pieces. An initial problem relates to the fairness of privatization as a solution. This concern is a bit removed from our focus, for it deals strictly with fairness among humans, but it is something that must be faced squarely before we start slicing up more of the land. Once our scarce resources are divided and allocated, the communal stock is gone; those who arrive later are out of luck. They must deal with and be at the mercy of those who arrived in time to get a share. Future generations, of course, are most plainly short-changed, except as the accidents of birth give rise to inherited privilege. Privatization divides the haves from the have-nots and means that, after it occurs, everything must be bought. It adds even greater truth to C. S. Lewis's remark, quoted at the beginning of this chapter, that power over nature means the power of some people over other people with nature as the tool.

A more pertinent problem with the common-pool image as a solution is that it assumes that individual owners will not destroy the property that they privately own. Unfortunately, we regularly confront evidence that they do, and I see some of the evidence as I pace my Illinois fields. When miners bulldoze hillsides to get at a few bags of gold, they do not mean to destroy their privately owned lands, but they do. With ever bigger machines, the farmer cuts open his soil every spring, knowing that it will gradually wash away. The timber company engages in clear-cutting practices that destroy the soil, not unlike the factory that injects its poison deep into the ground that it owns. Historic structures come down; tenements fall into disrepair; acid rain eats at our religious and secular cathedrals. Many of these cases are exacerbated in the common-pool setting, but the destruction occurs as well when the costs are all borne privately. In purely economic calculations, the future simply weighs too little to offer protection from today's human wants.

The willingness of humans to destroy what they own gets at the basic drawback of privatization as a solution to our environmental mess, however much privatization may help in mitigating some of today's common-pool disasters. Privatization will work only if users develop some sort of recognition that ownership rights are limited; only if owners understand that they must protect and preserve what they own, as well as use it. The simple story of the common covers up this truth, for the story suggests that economic self-interest on its own will provide a sufficient incentive to protect the Earth. In fact, we could privatize everything and destruction would continue, so long as the owners in charge lacked an ethic of care.

But let us delve a bit deeper and watch this common-pool story in action, accepting the various assumptions about the economist's rational actor. In the case of the common pool, economists suggest, the need is to regulate the pool so that it remains productive over the long run. To do this, we must first decide whether we want the pool to produce forever or whether we are content with production over some set period (ten years, fifty years, or whatever), at the end of which the pool is exhausted. In the case of the fisheries, we must decide the issue up front, for it will affect how we determine the size and number of allowable quotas. In the case of grazing lands that we divide into discrete plots, the issue is one for each plot owner to decide.

The difficulty with privatization of the fisheries is that the fishermen who get together and allocate quotas have no self-interest in keeping the principal (that is, the breeding stock) intact in perpetuity. They might well do so (and in practice sometimes do), but they will protect the breeding stock—the golden goose—only if they look beyond their economic interests and show concern for the Earth itself or for future generations. A truly self-interested fisherman has no need to worry himself with the pool after his death and will quickly agree to a harvesting schedule that causes the pool to be exhausted right after his projected date of death. If this schedule is kept, the fishery is soon depleted, and the Earth and future generations are the losers. Even if every fisherman wants a full-lifetime amortization schedule, and successive groups regularly postpone the ending date so that it is always a full lifetime away, this merely delays the inevitable exhaustion as the annual consumption is taken from a slowly reducing pool.

Even this much foresight is by no means assured from our self-interested fishermen, for many things can happen. For one, rule making can be taken over by septuagenarians whose horizon is close and who might well switch to a ten-year amortization period. Second, the group might include a few people who have trouble with the notion of saving for the future and who vote to cut into the principal now. Third, times might get tough and the group might cut even deeper, grasping the principal (like many of our farmers and loggers) while knowing that they or their children will later pay.

The point of the story is this: if people look only after themselves, they will cut into the principal, and they will cut in even more quickly if older or less cautious users are in charge. In the case of the grazing lands that we divide into discrete individual lots, the same sad concluding point is reached, only much more directly. Each self-interested farmer takes a grazing lot and uses it for his lifetime. His grazing pattern maximizes return during his life and leaves the land more or less ruined on his death.

The truth is, if economic self-interest is really going to be the sole guide, we cannot solve the common-pool problem simply by letting each owner, or even each generation, look after itself. In the long run, both shared-use and individually owned assets are susceptible to exhaustion by the present generation. Dividing common-pool assets among individual owners at best reduces the pace at which exhaustion occurs.

We can protect the Earth and provide for the future only if each generation believes that it ought to pass on as much and as good as it received—only if each generation embraces some type of ethic of care. This kind of caring protection, to be sure, happens in many instances, but it happens because people care about the Earth and about future generations—which is to say that they don't fit the mold of the economists' self-centered actor. Protection happens, thankfully, because the cares of many people go beyond the self-centered and the economic, and because people often act upon their self-less urges.

When we get down to it, the entire Earth is a common pool, a community of life of which we are a part. And in important ways we cannot divide the Earth into individual shares, even if we could do so fairly. When we divide the Earth artificially, we simply create smaller common pools. Truly self-centered owners will always want to cut into the principal, whether they own it individually or share it with others. And this action will always hurt those who come later, if only by depriving them of options that once were available.

When an ethic of care reigns, the land will be well tended; when it is absent, the land will lose, regardless of the ownership scheme.

A final, perhaps more subtle concern with the economists' common-pool story is that it suggests we can solve our environmental problem without tampering with our fondness for greed. Indeed, the story goes further and almost tells us that greed is entirely respectable and that we can solve our problems with greed as the main guide. As we shall see later, we have good reason to suspect otherwise, and we have good reason today to balk at the hint that all people now are dominated by greed. Privatization interferes with our sense of community, with our sense that we share the Earth with all other humans, if not with other life forms. It encourages us to think that some part of the Earth can be ours to do with as we will, without concern for the interests of others.

Externalities

The third economic conception of our environmental plight depicts our difficulties largely in terms of what economists call externalities. An externality arises when a person or business engages in activities that impose harms or benefits on others with no compensation paid to equalize the situation. When a factory dumps chemicals in a stream, the cost of the pollution is suffered by the people downstream, who are external to the factory and who are not compensated for their injuries.

What externalities thinking tells us, in simplified form, is that uncompensated harms skew calculations of profit and loss and therefore lead to the overproduction of the pollution-causing goods and services. Because the factory owner who pollutes need not pay for the pollution-induced harm, the owner can ignore that cost in calculating profits and in setting production levels. The factory in theory will increase production until its marginal cost of extra production equals the marginal revenue that it receives from selling the product. Excluding the cost of pollution understates true cost, making more production seem justified.

Like the common-pool image, externalities analysis comes with a matched solution: internalization. When a factory owner is forced to pay for the costs of pollution (through taxation, pollution fees, or liability rules), the external harms will be taken into account, or internalized. The factory's profit-and-loss statement will now reflect true costs, and production will be at the optimum (that is, the economically efficient) level.

Like the common-pool image, the image of the externalities-generating factory helps explain why polluting the Earth can prove so profitable. It helps show how actions on one acre stretch out and affect others, sometimes on the far side of the globe. By performing this role, the externalities image serves as a helpful tool.

There are several practical problems, however, with an externalities-based explanation of our environmental plight and with the proposed solution of internalization. Underlying many discussions of externalities is the implicit suggestion that the externality-generating case is atypical, easy to spot, and easy to remedy. The usual hypothetical example involves facts much like those just used: a factory that harms identifiable nearby landowners with some particular type of easily traced pollution. This hypothetical case describes one of the few occasions when we perhaps can remedy with after-the-fact compensation, which no doubt accounts for its popularity as an illustration of the basic idea: because the case involves a single pollutant with immediate, identifiable, localized effects, internalization seems feasible.

The troubles begin when we move from this easy case to the kinds of complexities and vagaries that nag us in real life, both on Main Street and on the back forty. First of all, the common case out in the world will involve pollution whose effects are widespread, hard-to-trace, and often long delayed—even perhaps multiple pollutants that interact bizarrely with an ecosystem. Externalities will often be horridly complex, testing the skills, tools, and patience of even the most diligent chemist. Second, externalities are far more common than our simple factory image suggests, for our every interaction with the Earth gives rise to them. The production of externalities, in fact, is a near-universal norm. Third, as we examine our actions closely, we soon start having troubles distinguishing between external harms—which we want to halt—and other external effects that might well be beneficial. Externalities in the real world don't come with white and black hats. And the more numerous, complex, and ambiguous they are, the more impossible they are to internalize.

When we cast a strange chemical into the sky or the ground, we typically have little idea what effects it will have. We did not know our air pollutants would break down the ozone layer and, by trapping certain light waves, contribute to the global warming of the greenhouse effect, yet that harm is now worldwide and, as best we know, more or less permanent. When we burn a gallon of gas or dump refuse in the ocean, we certainly cause external harm; the difficulty is that we don't know how or when the harm will occur. Thousands of deep-injection wells add tons of toxic wastes into the ground daily, and sooner or later, we know, groundwater and surface water supplies will suffer. What we don't know is when, where, and how. And then, of course, there are the time bombs of cancer, infertility, and other ailments that take years to show up and that typically can be traced only long after the fact.

Even greater difficulties arise when we go beyond pollution and try to use the externalities model to deal with resource-use issues. Each gallon of gas that is consumed reduces the Earth's stock, imposing indirect costs on every person in the world and on generations not yet born. Externalities arise when the local bottling company switches from returnable glass bottles to nonreturnable ones or from nonreturnable glass to laminated, nonrecyclable plastic. They also arise through inaction, as when a city's only newspaper refuses to use recycled newsprint. The newspaper's refusal to buy undercuts the market for recycled paper, causing financial troubles for the local recycling company and perhaps leaving residents no option but to toss newsprint in landfills.

Many of these complexities can be illustrated by considering the troubling case of the Earth's dwindling wetlands, which are drained and filled in preparation for new construction. The owner of the wetlands can claim that no externalities are generated by the draining and filling, for no polluting chemicals or materials are pushed into the air, dumped into the ground, or cast adrift in the waters. And if by externalities we mean traceable pollutants, the owner could well be right.

But, to the ecologist, destroying the wetlands has grave effects that spread widely, whether or not we decide to call them externalities. Wetlands play important roles in cleaning water and regulating water flows. They are the Earth's kidneys, and their cleansing role is irreplaceable. Wetlands are prime spawning grounds for fish and vital habitat for many birds. They are ecosystems unto themselves, full of diverse and threatened life forms. When a single wetland is filled, the costs are small and perhaps almost unnoticeable; as more wetlands are filled, however, the costs rapidly escalate.

Flooding is harder to contain, and engineering teams must spend great sums dealing with the problem artificially. Surface waters begin to retain more pollution, and plant and animal species rapidly suffer.

What is true of the wetlands is true as well of all land, if less visibly so. Whenever an open acre of natural life is converted to a parking lot, nonhuman life suffers. Every open acre supports wildlife and adds to the Earth's richness, vigor, and diversity. The western rancher does not act benignly when he erects a fence that halts a pronghorn antelope migration and leads to mass starvation. Like the wetlands developer, the rancher can claim that his actions impose no pollution on others. But the issue, needless to say, is not so simple.

This confusion about what is and what is not an externality stems in part from our system of property rights and laws and from the difficulty of deciding what constitutes a harm. If we say wetlands owners have the right to develop their land, the inevitable losses that result may well go overlooked. If we say owners have the right to drain and fill, we may conclude that no legally cognizable harm has occurred.

But saying that owners have these rights largely decides the whole issue before the conversation has even begun. As we deal with our environmental problems, our old definitions of property rights are going to need some changing. Once we start to erase the old lines in preparation for drawing new ones, the wetlands case becomes intractable. We cannot automatically identify what is external and what is not. We are at a loss to figure out what is a harm and what is not. To make these determinations, we first must make some important judgments about what type of conduct is right and fair. We need a scale to use in separating the sheep from the goats. And at this task, at the task of sensing the good and the bad, externalities thinking—and economic reasoning in general—offers little aid.

As we have seen, economists suggest that we can solve the problem of externalities by requiring the harm generator to compensate those who are harmed. Internalization, they say, should be the focus of our environmental laws. The underlying rationale is that a harm generator will pay only if the generating activity is sufficiently profitable and if the costs of payment are less than the costs of compensating for the harm. Payment indicates that the harm–generating activity is economically productive and, hence, desirable.

This reasoning, however, is based on assumptions that are naggingly incomplete and factually problematic. Internalization makes sense only if we can identify the harms—which we often cannot do—and if we have feasible mechanisms for arranging compensation—which we do not have, given the enormous costs and delays of large–scale litigation. An even graver difficulty is that the harms are often irreversible: once done they are done, and the world is thereafter a worse place. If all we demand is cash payment for environmental harm and call that a full remedy, we give the polluter the power to engage in the most cruel form of resource acquisition. Environmental loss is not corrected by paying money; our soils, our wildlife, our water and air cannot use it to heal themselves. Only when the threat of liability acts as a deterrent can internalization help. After–the–fact payments simply make destruction a cost of doing business. Our Earth, then, is truly for sale.

When calculating payments to those harmed by pollution, economists would have us look only to the damages of living humans. When we do this, however, we greatly understate the harms. An oil spill in the ocean can cause grave harm to nature, but the one who caused the spill may sail away, scot–free, unless those who fish the area can show a decline in their take. Our laws must go much further and protect all of nature, and only small steps have so far been taken.

By remedying only identifiable harm to people, we ignore everything that we cannot sense, cannot trace, cannot value, and cannot wait to experience. We also ignore the interests of generations to come, since compensation paid today provides them with nothing. We have no way of knowing, for example, what benefits might have come from the many species that we exterminate each day. This problem of undervaluation (or nonvaluation) becomes more extreme when we depart from our anthropocentrism and recognize intrinsic value in other life forms and ecosystems. These harms, too, will go undervalued unless and until we somehow add to our externalities analysis a new, far broader definition of harm.

The externalities image carries simplistic messages that make it harder for us to come to grips with our communal problem. Externalities analysis is based largely on the idea that pollution is the source of our problem. But pollution is only one symptom of a deeper discontinuity in our dealings with the Earth. It is the most obvious manifestation of a larger problem—the fever that provides evidence of infection.

If we look out into the world honestly—that is, free of old conceptions about economic entitlements and property rights—externalities analysis begins to seem artificial, whatever its theoretical value. It is too simplistic except for dealing with the most obvious pollution cases. To make the theory accurate, we must factor in an express recognition that we do not know the harms, we cannot trace or value them, and in a real sense we cannot remedy them. Like other species, humans cannot live without changing their environment, and the biosphere in which we live is itself constantly changing. We can end all externalities only by disappearing from the Earth. External effects are inevitable, and somehow we need to distinguish effects that are harmful and unacceptable from those that are not. To do this, however, we must go beyond externalities theory to something that helps us differentiate sustainable practices from ones that we cannot sustain.

The Unfettered Market

Both the common-pool and externalities conceptions represent attempts to correct errors in the way that the market establishes prices and guides the use and consumption of parts of nature. The idea behind both approaches is that a perfectly functioning market would assign to each component of nature an accurate price relative to all other goods and services. Market transactions are undertaken by people, not, of course, by other animals or inanimate nature, so it is inapt to view a perfectly functioning market as an image of nature. Still, the free market is for many observers the most attractive answer to our environmental problems, particularly if the market can be cured of some of its defects through privatization and internalization.

Perhaps we can think of the free market as a type of human community living on and interacting with nature, grabbing pieces of the land, trading them and using them, and through countless individual transactions determining the proper value of each piece. When the market works properly, each part of nature will receive its rightful price and will be valued and respected to the extent of that price. According to this line of thinking, when every part of nature is properly valued, the Earth as a whole will receive its due respect, and we will be dealing with the land in the right manner.

This line of argument is sometimes presented in a different way. The unfettered market, proponents point out (and rightfully so), is fully consistent with a stern environmental ethic. Nothing stops an owner of resources from using them in a respectful, sustainable manner. If all owners adopted a responsible ethic, the market could still flourish and could, in the most efficient mode, allocate resources to the highest-paying, ethically permissible use on a one-dollar/one-vote basis. This mechanism of bringing preferences together works every bit as well among ethical consumers as it does among exploiters.

This reasoning seems to lead many observers to proclaim, in essence: Hands off the free market. Change individual preferences, but leave the market alone. Or, perhaps more accurately, correct the common-pool and externalities problems, as best this can be done, and then leave the free market alone.

There are several responses to be made to these overlapping lines of argument. First, in the free market, the ethical and the unethical are treated with equal respect. Each has a voice equal to the dollars that it commands. It is true, as economists say, that a free market would work fine in a world of Earth-sensitive people, but it would also work fine in a world of dedicated Earth abusers. The free market will not stand in the way of people who are inclined to foul and consume the irreplaceable. The virtue of the free market is efficiency, with efficiency defined in accordance with the desires of every person with dollars to spend. For good or ill, it is neutral, in the best democratic, libertarian tradition. The market, that is, is not the cause of our problem, but it is also not the cure—even when thoroughly cleansed of its imperfections—and this is the main point.

The market, then, is value free, yet our concern is precisely the topic of values—what they should be and, more particularly, how we should express them in summary form. In this discussion, economists will need to back away from center stage. The Earth-centered values that we have been and will be considering are, most fundamentally, personal values, the values that will shape our preferences and our judgments, both political and economic. As the public discourse over values continues, people's preferences will change. At some point, or so we can hope, each of us will settle on a moral scheme, and perhaps the free market will then be the best way to bring our individual preferences together to form group decisions. Perhaps, in some settings, it will not be the best. But whatever we ultimately decide about the market and its role in allocating goods, one idea needs to be clear in our minds: the issue of values comes first.

Two other responses can be made to free-market advocates, both more accusatory. In operation, the free market is not as neutral as it seems. The undercurrent of economic thinking—in its popular, nontechnical, most influential mode—is of the "I'm OK, you're OK" mode: what I do is fine, what you do is fine, and somehow the market will take care of any problems. In fact, the market does not clean or cure; it merely translates. If we value the Earth very little, the market will facilitate abuses, plain and simple. In subtle, perhaps largely unintended ways, economists tell us to have faith in the knife. But the knife alone does nothing. It is the one who wields the knife who counts—be it the healing surgeon or the careless butcher. By focusing on the knife, we ignore the knife bearer's soul, which is where we find both the problem and the solution.

A more pointed response to free-market advocates is that, by promoting an unbridled market, they help undercut the governmental process, which provides one of our prime methods of discussing public values. When people come together to engage in lawmaking—as voters, interest groups, party activists, litigants, or whatever—they do more than make laws. They engage in a complicated conversation over values. Perhaps no values are more public, or need to be more public, than those that deal with the Earth–humankind interaction. As a community, we need to talk about these issues, and one important way to do so is for us to gather as a community to set rules for communal conduct. The thrashing and debating that goes on in such gatherings is a vital educational process, for all concerned. It is probably no coincidence that the more public and open we are in setting our rules, the more ethical the rules are likely to be.

Conservative free-market enthusiasts—largely unintentionally, we can assume—would cut off this debate and thereby frustrate an important growth process. In popular free-market thinking, landowners like me should be able to do what they want on their own land. Land-use rules interfere with this freedom and so are bad. If rules are bad, there is no reason to talk about their shape and content: they simply should not exist. In its free-market form, economics has no vision of a moral good.

These complaints about free-market thinking are not

confined to the setting of humans in nature. Economists take people as consumers, and consumers are people who set their values and consumptive preferences (with the help of Madison Avenue) before they enter the market. The interaction in the marketplace is the abbreviated, self-serving communication that occurs between individual buyers and sellers. No mechanism exists for consumers to get together and, as a people and a community, talk about what they want. By allowing the free market to impede public discourse, we hamper the growth of a public, shared, and more lasting morality.

———

In my thinking about how to use my woods, creek, and fields, I am aided only a little by what economists have to say, and I am, frankly, frustrated and annoyed at the apparent popularity of econospeak. I know my land is not permanent—the evidence of that fact is plentiful on my land alone. Because I am sole owner, the common-pool story doesn't help me decide what and when to plow and plant. At most it encourages me to know that my time on Earth is short—a fact that, again, I can and do learn in other ways. If the county records list me as the only current owner, they tell me also who came before and, by extrapolation, that others in turn will follow. With the earlier owners, as with those to come, I use the land as a common home. My fair allocation is to use the land while I live and then to pass it along, as healthy and productive as before, when the next owner's time begins.

I have my hardest time trying to figure out what to do with the economists' externalities image and its recommendation that I internalize. The truth that is contained in this message is the truth that the overflying geese proclaim in their semiannual migrations: my acres are linked to all else on Earth. But I would have it no other way, and I'm not about to erect walls at my borders or buy a computer to help me enumerate and calculate all that comes and goes across my edges.

With these economic messages in hand I also feel no better able to translate my land into a dollar price. I read, and think, and read some more, and I still cannot comfortably translate my land into a number on a purchase contract or a balance sheet. I still am at a loss on pricing the red-tailed hawks, the flickers, and the meadow voles that have their competing deeds to my land. Dollars, it seems to me, are for people, and dollar numbers respond to transient, fluctuating, human concerns. Am I wrong to sense, as I do, that the land somehow is above this pettiness of prices, that it has, or should have, a stability of value that dollars can never have?

The creek that crosses my property goes by several names—Hog Branch, Hog's Branch, Hogue Branch—depending on the source that I consult. Some two or three hundred yards past the point where the creek exits my property it flows into the Embarras River, a tributary of the Wabash, a tributary of the Ohio, a tributary of the Mississippi. *Embarras* is the French word for "trouble" or "obstacle," and early French explorers and missionaries attached the name to more than one water flow in the Midwest. Each was named, not for any natural attribute of its own, but because of its initial effects on the first European visitors. I can only guess why my local Embarras aroused some traveler-mapmaker's annoyance, but, in my neighborhood, it is an obstacle in one obvious sense, at least for humans. There are no nearby bridges, which is why traffic by my land is rare. My land, in fact, lies in a large bend in the river, along with a few other fields. To the traveler not intending to stay and not going where the river wanted to go, this Embarras, like all the others, was simply a block in the path.

Like the early French travelers, we evaluate the land that surrounds us based on its utility or disutility for our immediate needs as we perceive them. What we don't want, we manipulate or discard, and we assign it a low price. When we assign an object a price, we equate its value with other objects of the same price. Along with low value comes degradation and decline.

3
Owning the Land

Property belongs to a family of words that, if we can free them
from the denigration that shallow politics and social fashion
have imposed on them, are the words, the ideas, that govern
our connections with the world and with one another: property, proper, appropriate, propriety.

—Wendell Berry

At the southern tip of New Jersey lies Cape May, the northern terminus for a ferry that transports travelers across the
wide mouth of Delaware Bay. William Penn's colonists passed
this way more than three centuries ago as they journeyed up
the Delaware River to found the city of brotherly love. Today,
naturalists from around the globe arrive here, binoculars and
spotting scopes in hand, to gaze at the still-dense flocks of
birds that migrate along the coast.

Beginning just east of Cape May and extending along the
Atlantic Coast of New Jersey lies a string of barrier islands,
which are laced, particularly on the leeward side, with fertile
saltwater marshes. The marshes are rich in cordgrass, salt
meadow hay, needlerush, and sea lavender. They are visited
by kingfishers, great blue herons, and the migratory clapper
rail, the bird that naturalist George Reiger has called the

marsh's supreme symbol. Black ducks come here, as do sandbar sharks and immense numbers of fish and invertebrates.

Barrier islands like those along New Jersey are, by the standards of geological time, perhaps the most transient of all land forms. As their name implies, islands like these absorb the shock of the ocean's ceaseless waves and are battered and misshapen by ocean storms. In the course of a single human lifetime, barrier islands can move, shift, and take on entirely new forms. There is no end here to the mapmaker's work.

Although the impermanence of barrier islands has long been known, this knowledge only recently has slowed the rush to build beachfront homes and hotels. Extensive building projects mark New Jersey's islands, projects that will require, in decades to come, increasingly expensive efforts to defend them against the undeniable forces of nature.

On many of these barrier islands, developers have built on saltwater marshes that they have drained and filled to create solid ground. Government money has helped fund the process. Wetlands draining has long been a national crusade, and more than half of what our nation once had are gone. Not until the modern environmental age was there any widespread acknowledgement that the filling of a wetland involves a cost. Today, wetlands are the prize in the war between those who would continue conquering and altering the land and those who sense that our work with bulldozers and drainage tiles has begun to yield, not just diminishing returns, but negative ones.

One of the battlers over wetlands is Robert Ciampitti, a man who has succeeded in altering large portions of Diamond Beach on New Jersey's southernmost barrier island. Ciampitti is a real estate developer, and he is good at it. Like many developers, he enjoys seeing natural areas turned into building projects—particularly natural areas that he owns. He is a modern crusader in the quest to subdue the land, a pioneer in a business suit.

Ciampitti began his work at Diamond Beach around 1980. From the beginning he succeeded in buying low and selling high. As his operations grew, he bought larger parcels and more marginal land throughout Diamond Beach. In September 1983, Ciampitti acquired forty-five acres of undeveloped land for $3.3 million. It was this purchase that would cause Ciampitti to stumble and to spend years in the courts, for included in the purchase were fourteen acres of state-designated, state-protected wetlands.

When Ciampitti bought his wetland acres, he took on the mantle of private owner, perhaps the most exalted garb in the Anglo-American system of common-law jurisprudence. Private property, particularly land, is something that we go to great lengths to protect. Ciampitti bought his wetlands intending to turn low-lying marsh into blocks of private homes. What he didn't count on was the combined opposition of the New Jersey Department of Environmental Protection and the United States Army Corps of Engineers. By law, both agencies needed to grant permits. As he soon learned, neither wanted to do so.

Robert Ciampitti, however, was not a man to be easily turned down, and he quickly shifted his fight to the courts. As he saw the situation, a permit denial was a form of economic death. Ciampitti assembled a team of lawyers, engineers, and expert appraisers. Together they challenged head-on the constitutional power of the government to deprive him of his development rights. Ciampitti's ace in the hole was his expert appraiser, whose testimony seemed startlingly clear: With full rights to develop, Ciampitti's wetlands were worth over $300,000 per acre. Without the right to build, the value per acre was a mere $200—a decline of over 99.9 percent. Surely, Ciampitti would argue, his land had been taken from him for all practical purposes when the government denied his permit. Surely the taxpayers should pay for this multimillion-dollar loss.

Ciampitti's legal claim would spend years winding through the courts. As it did, developers from coast to coast found themselves in the same bind, and many would join in resisting. Stripped of its legal veneer, Ciampitti's claim was simple: if the government wants to protect wetlands or any other land, it should buy them. When government regulation compels a landowner to leave land vacant, it interferes with the landowner's inherent right to develop, and this it cannot do. A limitation on development is one thing; an outright ban is quite another.

Claims like these have been hard for our legal system to handle, for the assertions on both sides are powerful. Our long-held sense is that society's economic and political stability depends on reliable property rights. If landowners cannot reap where they have sown, prosperity will decline. Yet as ecologists tell in vivid detail, wetlands are vital, irreplaceable ecosystems, and their destruction brings grave costs. Wetlands offer erosion and flood control, water purification, and habitat for fish spawning and wildlife reproduction. When wetlands are destroyed, everyone suffers—except for the landowner, who typically captures most of the resulting gain.

When we think about the claims of landowners like Robert Ciampitti, our minds naturally draw upon images of land ownership that we obtain indirectly from our unique legal culture. When developers claim that they have the right to fill marshes, we instinctively evaluate their claims by turning to the meaning of land ownership as we grasp it. Has the government, we wonder, interfered too seriously with the core attributes of what property is all about? Has the government essentially taken an owner's land?

Lawyers and lawmakers by and large are a conservative lot, in the language and ideas they use as much as in the clothes they wear. The law has deep roots in the past, stretching back to Anglo-Saxon England in the tenth and eleventh centuries. Lawyers are fond of digging up old precedents and carrying forward antiquated words and phrases. The law changes, at times even rapidly, but change is often made by pressing new ideas into old forms to create an appearance of continuity. Old understandings linger on, even when they poorly accommodate new realities.

From lawyers and from courtrooms these legal visions permeate society and help guide the ways in which we think instinctively about the physical world around us. They furnish part of the backdrop against which we react to environmental problems. Our task here is to dig these legal messages and conceptions out of their time–encrusted formations, to grasp how they influence our understandings of how we ought to relate to the land. After doing this, we can return to Robert Ciampitti's tale.

Nature as Property

When lawyers refer to the physical world, to this field and that forest and the next–door city lot, they think and talk in terms of property and ownership. To the legal mind, the physical world is something that can be owned. It can be owned by individuals, by families, and by groups of people operating as businesses and nonprofit entities; it is something a community can own through its government. Ownership is preceded by a process of division, as the surface of the Earth is carved into distinct pieces before being allocated to owners in one way or another. Division is undertaken with great exactitude as surveyors draw lines calculated to the fraction of an inch.

When a person gains ownership of some parcel of land, that ownership includes all that is attached to the land—the grasses and trees, the nests and burrows. In the eastern part of the United States, land ownership carries the right to make use of rivers and lakes that the land touches. In the more arid western part of the country, water is a discrete asset that is allocated by separate permit to those who first put it to use. Those who own land also own the minerals beneath the surface and mineral rights can be (and often are) severed from the surface and sold separately. As the economy has become more complicated, even air rights have been severed from surface ownership and sold to the highest bidder.

Property has enjoyed a long and shining history in America. In medieval England, land was the principal form of wealth. Land ownership meant security, independence, and status. In modern times, other forms of wealth have overshadowed land in importance, but private land ownership lingers on as both a goal and a vaunted symbol. Land retains great value, and home ownership remains a badge of independence and success. Land represents more than wealth and financial stability; it provides a haven from the outside world, a place to escape from officiousness and intermeddling.

In past centuries, private ownership, particularly of land, was a central element in political debates. When Englishmen resisted the claims of royal power during the seventeenth-century civil war, they did so by asserting their property rights. Property was a source of strength to use in resisting the crown; it provided income and power and was thus a bulwark of protection against royal interference and an intrusive central government. When the American colonies resisted unwanted intrusions from England during the revolutionary era, they too viewed property as something more than wealth and status. Property ownership was an inherent human right, and private land was needed to counterbalance the power of the state. Widespread ownership, they believed, added stability to the civil state. In the years following independence, Thomas Jefferson and others talked forcefully about the need to promote land ownership among small farmers. Land ownership, they argued, gave people a stake in society; it made them more stable and sensible, less subject to political pressures.

In light of the importance that we attach to property ownership, it is hardly surprising that property law supplies us with some of our most foundational, most powerful ideas about the land and about our place on the land. So ingrained are many of the law's messages that we rarely give them thought. Yet these messages exert great influence on us when we make decisions about using the land and when we evaluate the rightness or wrongness of everyday land-use practices.

Our legal conception of ownership tells us, first, that the Earth is something that we can own; it is something that can belong to us and over which we can exercise the various rights and prerogatives of ownership.

It tells us, second, that the land can be divided into distinct, discrete parcels, and that division of the Earth in this manner is sensible. Once a dividing line gains the imprimatur of the law, it suddenly ceases to be arbitrary and artificial; it gains a certain soberness, something that we respect. The law's implicit message is that the physical world divides easily into component parts, with the water owned by A, the land by B, and the subsurface mineral rights by C.

The third component of this message is that our home is our castle, our zone of personal influence where *we* make the rules. This sense of security is protected in the Constitution, with its ban on entry onto private land by government agents without judicial warrants. We have the right to exclude, the law tells us, and we can enforce that right by invoking criminal and civil trespass laws. We are also protected by constitutional rule from having our property taken from us, even a small corner or portion of it, without full compensation.

If we merge these various ideas, we produce an image of a physical world divided into pieces and subject to private ownership and control, a countryside populated with castles, each with an owner who controls. This is perhaps the law's dominant image, a carryover from centuries past. It should be quite plain to us upon even brief reflection that this image of ownership and domination stands in the way of environmental progress. It suggests that we may do what we want on the land that we own, that what we do is no business of others, even if it involves destruction, and that what others do is no business of ours. The private-fiefdom image also extends to things other than land. An automobile is ours to do with as we will, and we can ruin, destroy, and cast off any item that we buy. To own anything, the law seems to suggest, is to dominate, control, and, if we want, to destroy. The implications of all this are disturbing, for they ignore even the most fundamental principles of natural health.

This private–fiefdom image arose under circumstances far different from those that we face today. It arose when people were either unable or unwilling to see the innumerable ties that make all of nature a seamless fabric. Decades ago, the law's principal tasks were simple: to get resources into the hands of people and determine the relative rights among them. Ownership norms rarely addressed the question of what things we should privately own.

Today, our natural resources are largely all allocated. Established rules crafted over the centuries protect the rights of the owner over all other people. Today's lawmaking task is much different and much harder: to determine how to reconfigure what it means to own so that we do not exceed the bounds of sustainable–use practices. In this task, law-makers are just beginning.

As ecologists have explained, often in delightful detail, the world around us is intricate and intertwined. Each part of nature is attached to each other part, and what we do to one part has effects that spread widely. When a developer like Robert Ciampitti fills in a saltwater marsh, the ripples spread far and wide. Fish populations decline, for most commercial fish species depend on wetlands for an essential part of their life cycles. Waterfowl decline even more seriously, for they too depend on wetlands for one or more critical needs.

If our legal culture is going to reflect this rich reality, if it is going to incorporate the numerous natural links that bind one acre to all others, ownership norms need to be based on a vision of property as community. The person who becomes owner of a tract must be seen, not as some recluse on an isolated island, but as part of a natural as well as social community, with all of the obligations that accompany that status. People can close their doors and retreat from the rest of humanity. But they cannot cut off their land from the rest of the Earth.

Our typical reaction to limits on property ownership is to suspect that someone else is gaining the upper hand. When our rights are limited, we almost instinctively assume that something is being taken from us—or stolen from us—for the benefit of someone else. This age-old response is one that we shall need to cut loose, for it has served us ill. It derives from an older time when property was simply a way of drawing boundaries among people. The boundary for our age to draw is the one that will separate people and nature, a boundary to protect our communal home—not between neighbor and neighbor, antagonist and antagonist, but between humans and the Earth.

Property still serves as a bulwark against an overreaching state, but the point we need to grasp is that it can serve this purpose even if we change our resource uses dramatically and treat nature with greater respect. We can keep out police officers with no warrants and still insist on wetlands preservation. We can demand compensation when land is taken and still come up with tough land-use restrictions that reflect an understanding of how acres and resources are linked. Resource rights can still be privately owned and freely traded; they simply must be defined in the package of entitlements that compose ownership so that the owner is stimulated to maintain a natural harmony.

Ownership schemes across cultures have come in a dazzling variety of shapes and forms; indeed, the concept of ownership means nothing in the abstract, stripped of a social context. Ownership norms are based on the shared understandings and needs of a community. When communal values come to change, when we imagine a way out of our environmental plight, our property norms will rightly and naturally change along with them. Until then, the legal battle over wetlands will continue to surge aimlessly.

Property as Civil Rights

If we continue to probe the law's implicit messages, we find that property law as lawyers describe it has little to do with objects, or what most people colloquially call property. To the lawyer, property laws do not deal with the rights that we have in our house or bicycle. They deal instead with our rights as against other people's with respect to our things. My rights as homeowner are the rights that I have against the whole world relating to my home. My right to exclusive possession, lawyers tell us, is really my right to insist that other people stay away. As owner, I have the right to use, manage, alter, transfer, or destroy as I see fit—so long as I respect the similar rights of other owners—and these rights are all ones that I have and that the rest of the world does not.

This sense of property as dealing with rights among people has often been helpful. When property norms focus on people, the object under dispute fades from the picture. We can consider more abstractly whether, for example, an owner of property should be able to make a gift of the property, effective upon death, without complying with the requirements of a valid will. This is the kind of abstract issue that property law considers, and for most of this work the details of the object are unimportant.

The problem with this conception of ownership lies in its abstraction. By focusing on people, it diverts attention from the fact that not all things are alike, sometimes not even remotely alike. Abstraction makes it harder to develop ownership norms that take into account the peculiarities of the object being talked about.

We can ask abstractly, for example, whether the owner of land has the right to build on it or drain a wet spot or erect a fence. When the focus is entirely on the owner and other people, we are inclined to render an answer that is uniform with respect to a various land parcels. When our focus instead is on the land parcel and its peculiar features, when we assess the ecosystem of which it is a part and the animals and plants that make use of the spot, the answers will vary, perhaps widely. One owner may be allowed to drain or build because the effects are slight; another may not have these rights because the damage would be too great.

We need to resist the legal image of property rights as simply a question of rights among people, because, in fact, far more is at stake than the demands of today's human participants. In coming years, property norms will need to be based on context, on accommodation to the needs of all surrounding life. Our vision of what it means to own must include the rather commonsense notion that property norms deal with concrete things. When we keep in mind the thing, when we look closely at its peculiarities, we are more likely to come up with ownership norms that take these peculiarities into account. An acre of saltwater marsh is not the same as an acre of fast land on a barrier island, and neither is the same as an acre of pasture on the mainland. Why should our legal images treat them like triplets?

Who knows when and how we got onto a simplifying track so insensitive to the differences among the items of property we own? Who knows how we came to equate the erodible hillside with the prime flat field? Perhaps we were swayed by the powerful forces of industrialization, which encouraged people to think in terms of interchangeable parts and fungible inputs and outputs. Like the laborer on the assembly line, land somehow became an input, faceless and tool-like, and the Earth has suffered as a result. Whatever the causes, it should be clear today that we need to regain a more sensitive vision.

A further problem with the idea that property norms deal only with the relative rights of people is its tendency to focus attention entirely on the loss that humans bear when the land is damaged. When a parcel of land suffers an injury—when soil is eroded, trees burn, or birds are poisoned—our legal culture tells us that the injury is really suffered by the land's current owner, not by the land itself. This injury, moreover, is equal in amount to the landowner's provable financial loss. If the land is injured in a way that the market does not value, the injury is irrelevant. Since most plants and animals have little market value, the category of irrelevant losses is vast. By implication, if the owner is reimbursed for the financial loss, the injury is remedied and the case is closed. When the upwind farmer spreads the nerve gas parathion on some wheat fields, part of it will drift with the wind to the adjacent marsh, bringing death to many creatures. With the help of the law's implicit guidance, it becomes easier to overlook this loss, and for a long time we've done so.

This same problem of abstraction existed in early America with respect to slavery, and a glance at history may prove instructive, if not chilling. In the slave codes, a person who killed a slave owned by another was obligated to pay for the loss. The loss was measured in monetary terms; when payment was made, the injury was remedied. If the slave was too old or sick to work and could not be sold, no loss occurred. The deceased slave had no claim, nor did the slave's family and friends.

The slave codes operated as they did because slaves were a form of property, and property norms dealt, then as now, with the rights and interests of the owner. If we are serious about improving relations with nature, we must stop thinking of it as a slave.

The Law's Language

In addressing the topic of justice and the Earth, we need words and concepts to express our meanings. If words were mere tools that we used as we chose, we would hardly have any reason to comment on this obvious need. But words are not simply tools, and the task of developing a shared vocabulary is both more tricky and more important than first meets the eye.

Words help shape the world and the ways in which we are able to interpret it and think about it. Our ability to develop thoughts and to express ideas depends on our access to the right words or phrases, and the task is much harder if we don't have them. The words that jump into our minds, the ones that seem most available for use, direct our thinking more than we know.

When we think and talk about right and wrong, we use a vocabulary that largely comes out of the law. It is rich in dichotomies. Party A owns a certain piece of land and has the right to keep you off it, the law says, and you have a duty to stay off. B has the privilege to do something, and you have no right to interfere. C has the power to tell you to do something, and you are under a liability to obey. Rights, privileges, and powers, however precisely we use these terms, are interests that are legally enforced and that make sense only if they can be enforced. A right makes sense only when enforced against some other actor, someone who has the ability to obey or refuse.

When we try to use this vocabulary in dealing with nature and future generations, the words don't seem to fit. How sensible is it for me to stand in front of a redwood and proclaim loudly, to the tree, that I have the right to cut it down? How sensible is it for me to proclaim loudly to future generations that I have the right to use all of the oil on Earth and leave none behind? Somehow, surely, this game is not fair.

If we have rights against trees and future generations, then they must have corresponding duties, but how can they have duties when they are unaware of them? We experience the same sense of uneasiness when trying to talk about the rights of animals. If an animal has rights, those rights must somehow be like ours, and they cannot be taken away at will. But how can an animal have rights if it is unaware of them and cannot enforce them? If an animal has any rights, it certainly must have the right to live. That being so, how can we kill animals for food?

The big problem with this rights terminology is that it was developed for a much different purpose, and we are using it far out of context. Rights talk was developed to express the ways that one person should interact with another. Much of it was designed, in particular, for describing relationships between the citizen and the state. The entire discourse presupposes a context of person–against–person interaction involving actors who are roughly equal. In fact, it is sometimes difficult to use this language to describe situations involving humans who deviate from this unstated norm—the incompetent, the comatose, the fetus.

When we try to talk about rights *against* nature, or even the rights *of* nature, we use legal words in ways that are new. When we claim a right against nature or future generations, we are greeted with silence, and it is all too easy to interpret silence as consent. In some subconscious way the issue seems to be between us and the thing; if the thing offers no complaint, we resent outside interference. In fact, however, when we make such claims, we are not talking to nature, nor to future generations. We are, instead, talking to one another. And what we are talking about is not the issue of rights and duties, it is about nature, about other species, and about how the human animal *should* be conducting its affairs. The law's vocabulary of rights and duties makes little sense in this new context. Because of this, we need to use its terms with considerable caution—and we might do well to avoid them entirely, because, in the end, they seem likely to sow as much confusion as guidance.

If we find it helpful to use dichotomies, perhaps the best idea is stick to ones that are basic and unconfining, ones that do not channel or distort our understanding. We should simply talk about what is right and wrong, what is wise and foolish, or what is fair and unfair. If less precise, these words nonetheless are more likely to put our attention where we need it to be.

Freedom and Equality

We have seen so far that the law's implicit messages are largely hindrances in developing greater harmony with the

Earth, however useful they are in promoting fairness among people. This may seem surprising, for the law is flexible and contains many rules to protect the environment. In addition, elected representatives have wide powers to rewrite laws at will.

What we are trying to grasp here, however, is not the content and details of laws but the orientation that they provide—the ways they guide our thinking and get us subconsciously to understand how we fit into the world. Along these lines there is a final way in which our legal culture affects our thinking—perhaps its most important influence.

American political and legal culture places enormous weight on personal liberty, on the opportunity for all individuals to live as they want. Liberty is the central strand of the American fiber; it is what America is all about and why people flocked here from other lands. Equality is a comparably powerful beacon. Much of the civil rights crusade has been a quest to gain equal treatment, under law, for all individuals.

As ideals, freedom and equality go beyond the political realm to permeate all aspects of human life. What one person has the opportunity to do, we think, all should have the chance to do. Luck and ability, we know, play a good part in what a person is able to accomplish, but we resist the notion that the government should cut off opportunities. We resist the idea that our laws and our community can treat us differently from others. Fair treatment means maximum freedom from legal constraints on our ability to develop and express ourselves as we see fit.

In theory, there is no reason why we can't exalt freedom and equality in this way and still make progress in healing the environment. Freedom and equality are words that largely refer to the ways that people treat one another, and if we were to limit their meaning in this way they would cause little trouble. The risk is that we'll fail to confine them and that, in our enthusiasm for the values, we'll apply them in our dealings with nature.

If freedom is allowed to expand its reach, it can readily mean freedom to do with our land whatever we wish. And if maximum freedom remains the prime goal of our culture, then resource-use restrictions will seem out of place. Equality can mean that what you do on your land I should be able to do as well, for dissimilar treatment is unequal. If you can drain your pothole, I should have the same right, whether or not my land is similar in its natural attributes.

When we think more seriously about freedom and equality, we are likely to realize that neither of them is an end goal in and of itself. We seek freedom so that we can live as we wish; it is a step on the path to a good life. But surely a good life should not involve destroying nature, particularly when the ill effects are shared with others. Surely a healthy natural home is a shared substantive value that transcends our striving for individual choice.

Equality, too, is more of a tool than a goal. No two people are exactly alike. When we say we will treat people equally, what we mean more precisely is that we will deliberately ignore their differences. Issues of equal treatment typically arise in narrow settings, and in proclaiming equality we mean that, in some particular setting, certain differences among people should be irrelevant. When it comes to voting and civil rights, the race and sex of the voter are irrelevant, but the same is not true for age and citizenship.

In the environmental setting, the equality issue goes more or less like this: should we treat property owners equally when the property parcels that they own are different? If we are going to achieve harmony with the natural world, we are going to need to answer this question in the negative, and decidedly and loudly so. Part of harmonious living involves sensitivity to natural differences. It involves treating each parcel and each resource with respect, which in turn means recognizing its peculiarities. Urban zoning laws are a first step in this direction, but far more needs to be done.

Although equality is one of our highest goals, we very often feel that equal treatment is unfair. When people confined to wheelchairs ask for ramps and transportation equipment, they seek recognition of their unequal circumstances. Tax laws that assess taxes based on income recognize our differing abilities to pay. In fact, for every law that proclaims an old difference irrelevant, there is likely to be a new one that gives relevance to a distinction that once was overlooked.

Freedom and equality, in short, are ideas that we use to improve our lives, and we need to be free to use them as we see fit. In dealing with nature, we are likely by and large, to want to use other tools. To be sure, we still want and need norms based on fairness and justice. But fairness and justice in this setting are likely to mean—indeed, inevitably must mean—both restraints on how we live and resource-use restrictions that vary greatly from parcel to parcel and resource to resource, regardless of who owns them.

———

Given these various legal messages and understandings, it is easy to see why Robert Ciampitti was willing to gamble by buying wetlands on New Jersey's delicate barrier islands— and why his arguments elicit favorable responses from many listeners. In the law, Ciampitti's new land had distinct boundaries, just like other land; it was his new private refuge, his bulwark against the state and annoying neighbors, to do with as he wished. As landowner, his rights against other people were, by law, the same as any other landowner's. All he asked for was equality, for the same freedom to develop that other landowners had. His argument could touch all the right bases.

In January 1991, Ciampitti's quest for compensation encountered a snag. The United States Claims Court, Judge Bruggink presiding, announced that Ciampitti suffered no compensable taking of his property when the Corps of Engineers turned down his permit request. Several facts in the case influenced the court's decision. Ciampitti, it seems, was well aware of the wetlands restrictions when he bought his land, and he had good reason to know that development was impossible. His purchase price for the forty-five acres was fair even assigning the wetlands a zero value, and it appeared that Ciampitti had already recouped his investment by selling parts of the solid land. The court looked not just at the wetland acres but at the entire forty-five acres; doing this, it seemed clear that the Corps of Engineers had restricted only part of Ciampitti's land, and the parcel as a whole still retained great value. The government's action, in short, was not unfair.

Judge Bruggink's opinion gave official approval to a serious reduction in Robert Ciampitti's rights as landowner, but in doing so it did not use either new language or new images of what it means to own land. Judge Bruggink did not consider whether development of the land was consistent with the long-term natural health of the barrier islands. He did not ask whether landownership should carry duties as well as rights. He did not state that a landowner's rights are somehow dependent upon the type and nature of the land owned, nor that a landowner can use the land only in ways for which it is naturally well suited. In its language, Judge Bruggink's opinion is common; in its results, it is a portend of things to come.

In presenting his case to the court, Robert Ciampitti drew upon more than just the language of the law. He drew upon our long-standing fondness for almost everything called development. Development means growth; it means jobs, higher property prices, opportunities to earn money, and the like. An aura of goodwill surrounds this word like a halo. Resistance is possible, but the path is sharply uphill.

Favorable connotations like these add a heavy weight to one side of our balancing scales. Imagine how the preservation-versus-development debate would shift if we simply changed words: for *development*, let us substitute *destruction*. Both words refer to change. To develop is to build up; to destroy is to tear down. Wetlands developers means to develop, but to develop they must first destroy the ecological integrity and ecological virtues of a particular tract. Why do we call this two–step process development when half of what goes on, perhaps more than half, is destruction?

It would be safer, but no more (or less) accurate, to use the word *destruction* to cover both steps of this process. More fairly we might seize on words like *alteration* or *reconfiguration*—terms that do not pass judgment. The desirability of further alterations of our land urgently needs debating. It helps little to begin the talk with terms that pre-judge the issue.

———

In my home state of Illinois, some eighty–five percent of the original wetlands have been lost to the bulldozer and drainage line since settlement began less than two centuries ago. This figure is far above the national average, although below California's figure of ninety percent and Iowa's ninety-five percent. Most wetland conversions are undertaken for agricultural use rather than vacation homes. Nearly half of our national wetland losses have occurred in the seven midwestern states from Minnesota and Iowa to Ohio.

Wetlands protection laws have large gaps in them and are enforced by understaffed agencies. In northern Illinois, the Corps of Engineers is aided in its protection efforts by dentist Ken Stoffel, who spends nights and weekends leading a group called the Swamp Squad. Squad members, drawn from the Illinois chapter of the Sierra Club, seek out property owners who, unlike Robert Ciampitti, do not bother to seek permits. To date, they have caught nearly two dozen.

Recently, the United States Environmental Protection Agency awarded the Swamp Squad a $50,000 grant to continue its efforts.

4

The Moral Actor

In a gripping scene in Willa Cather's popular novel *My Antonia*, young Antonia relates to her friend Jim Burden the tale of two Russian men, Peter and Pavel. Decades earlier, when Peter and Pavel were young men still living in Russia, they served as groomsmen in the wedding of a friend. The wedding was in the dead of a harsh winter, and the large wedding party, after a long evening of partying, set out in seven horse-drawn sleighs to journey to the next town. As the wedding party traveled over the snow in the frigid, moonless night, with Peter and Pavel driving the lead sleigh, a pack of hundreds of wolves descended upon them. An accident overturned the last sleigh, and the occupants were jumped upon and killed by the hungry wolves. Accident followed accident, and one by one the wolves picked off the rear sleighs. As Antonia told the story, "The screams of the horses were more terrible to hear than the cries of the men and women."

When the lead sleigh finally approached the town the occupants, including the bride and groom, realized that theirs was the only sleigh remaining, and the wolves were closing in. One of the horses was tiring badly, and Pavel decided that their load needed lightening to outpace the hungry pack. Pavel sought to throw the bride out the back, but the groom resisted; in the ensuing scuffle, both bride and groom were sacrificed to the wolves. Only Peter and Pavel made it through alive.

Willa Cather based her story not on real life, not on true facts about wolves, but on an oil painting that she knew in her youth, a painting on display today in the Cather museum in her home town of Red Cloud, Nebraska. The painting is dark and menacing; the wolves hunger for the kill. It is easy to see how such a painting could stick in a girl's imagination.

Wolves were few in Cather's days on the Nebraska prairie of the 1880s and 1890s, and Cather would have had little familiarity with them. Most readers of *My Antonia*, initially published during the First World War, would have known little more. By then, the wolf had disappeared from most states. In Illinois, the wolf was gone by 1860, an easy target for bounty hunters and poison-laced bait.

Because Cather and her readers knew little about wolf behavior, it was all the more easy for her tale of Peter and Pavel to strike a believable, horrifying, nightmarish chord. Cather said what her readers tended to believe: wolves were dangerous, dastardly characters, ready to kill humans when conditions were right.

Wolves have long been feared and hated, and their extirpation was promoted by generous state bounties beginning with a Massachusetts bounty in 1630. Today, the gray wolf lives on in the lower forty-eight states only in the northern reaches of Minnesota, Wisconsin, and Michigan, with occasional animals drifting across the Canadian border into the northern Rockies.

Only in recent years have wildlife biologists and ethologists learned much about actual wolf behavior—to test and belie tales like Cather's, as well as the evil presumptions of generations of pioneers. Much of that research today is centered in northeastern Minnesota, in and around the small town of Ely, where the winters are as harsh as those on the Russian steppes and where horse-drawn sleighs remain a living memory.

What researchers have found, unsurprisingly, is that wolves are not gentle creatures—at least not to their competitors, enemies, and prey. Adult males can reach one hundred fifty pounds, and wolves operate fiercely and effectively in packs. In northern Minnesota, however, wolf packs average from two to eight family members—not the hundreds that Cather envisioned—and packs of over twenty-five are rare anywhere. Wolves have a strong dislike of humans, and they use their acute senses of hearing and smell to steer clear of people whenever they can. Documented wolf attacks on humans are rare; indeed, in recorded American history the only people ever killed by healthy (i.e., nonrabid) wolves have been Native Americans. Cather's story is a wild distortion.

In the Ely area wolves and people coexist, as they have done since the iron-ore boom brought miners into the region more than a century ago. Today, Ely's economy is largely based on tourism, and wolves and the more numerous black bears and moose add a mystique to the area that wilderness travelers find appealing. Ely is the main shove-off point for entry into the Boundary Waters Canoe Area Wilderness, a national treasure of clear lakes, forests, and canoe-camping dreams come true, a land of calling loons and shimmering northern lights. Visitors flock here in the summer—too many visitors, some wilderness protectors proclaim—to paddle, portage, and commune.

Even the rocks in this area carry a timeless air. Among geologists, Ely is known for the Ely greenstone formation, an exposed belt of basaltic lava—at three billion years of age, nearly the oldest rock found on the surface of the planet. The soil around Ely is thin—wiped clean by the last ice age—but it is fertile enough to support dense forests of pines, black spruces, balsams, birches, and aspens. The tallest of the white pines often harbor the weighty nests of bald eagles and osprey.

On the east end of Ely, as the visitor heads out the Fernberg Trail toward Moose and Snowbank lakes, one finds the International Wolf Center. Although environmentally oriented, its educational message is not sugarcoated. Its large stuffed wolves appear menacing. Photos show how wolf packs operate to bring down fleeing moose and deer. What the center wants to present are the facts, and the facts are that the wolf is a skilled hunter that more or less never harms humans. Wolves, to be sure, do kill livestock. But Minnesota deals with the problem adequately by compensating animal owners who present adequate proof of loss.

The growing talk these days about wolves and their habitat is mostly but not entirely symbolic. When wolves were eliminated, the top was lopped off the predatory pyramid. The animals that wolves kept in check by predation no longer faced their old foe. Their numbers in many settings rose, with humans largely left to fill the wolves' old job. Over the past century, the coyote—the prairie or brush wolf—has expanded its range widely to take advantage of the gray wolf's departure. Once a grassland creature, the coyote has adapted well to forests and has migrated to both coasts. Smaller and quicker than the gray wolf, the coyote thrives in human-dominated rural areas, which is why I have hopes that the coyote family on my Illinois land will not mind my intrusion.

Some wolf enthusiasts are motivated by the pleasure they get from living in an ecosystem that harbors predators large enough and mean enough to kill them. The thrill is there even though, in truth, death by bee sting is more likely. Most wolf enthusiasts, however, act less out of a need to seek thrills than out of a desire to respect nature—all of nature— and in that sense the wolf and its story serve to symbolize our overdomination. If we can live in peace with the wolf, if we can muster the restraint that such coexistence would require, perhaps we can in time show restraint toward all species and the Earth as a whole.

———

Wolf enthusiasts are but one part of a large, varied mix of people today who are intensely concerned about our dealings with other species, large and small. Included in this group are many university-based philosophers who think deeply about issues of ethics. Openly, sometimes boisterously, they are questioning whether our ethical norms should extend to include a few, or many, or perhaps even all other animal species. Should our dealings with nature, they ask, be based on ethical norms not unlike the norms that govern our dealings with other humans? Should we think of nature not as a collection of economic resources with price tags, not as a bundle of things that we can divide and reduce to private ownership, but as something possessing moral value? Should a gray wolf be a morally worthy subject and not a mere object?

Some of the philosophers thinking and talking about animals are allied with the animal rights advocates, or, as they sometimes style themselves, the animal liberationists or animal welfare advocates. Perhaps no group of nature protectors has attracted as much resistance and misunderstanding. Tree spikers seem overzealous but basically well intentioned in their efforts to protect old-growth forests; animal liberationists, by contrast, strike many bystanders as fundamentally misguided. They protest the blinding of rabbits to test a new color of eye shadow and complain of factory farming and the forced confinement of veal calves. Some talk of "liberating" zoo animals, of ending medical experimentation, and of seizing hunters' shells and anglers' hooks. We even hear proposals to sterilize herds of wild deer to prevent overpopulation. As the claims progress, they can appear increasingly extravagant, and the more extravagant, it seems, the more newsworthy.

Because of its high visibility, the animal welfare issue has largely overshadowed what is in fact a larger, more wide-ranging inquiry by philosophers into the ethics of our dealings with the Earth. Their concern has been not just with individual animals but with plants and nonliving matter as well as with species, ecosystems, and other entities above the level of the individual organism. We shall turn here to several of the images that philosophers have created as they have engaged in this debate, as they have talked about how we should live on the planet as thoughtful moral actors.

Insiders and Outsiders

When patriarch Noah built his ark, he assembled all the world's animals together for a grand and colorful parade up the plank. Two by two the species entered, the story goes, with equal place and equal status for all.

If we can reconceive today such a gaudy assembly of animals, we can begin to piece together the first image that arises from the philosophic literature. Let us bring the animals into one place and look them over, seeing where their similarities and differences might lie. Better yet, we can start with a larger group—all of the Earth's species—since plants need to enter the image at some point. In Noah's story, the species stayed together, and all were invited onto the life-saving vessel. In our modern version, not all species will be so lucky.

According to many philosophers, the key to our misguided stance toward nature is that we have improperly divided the species. We haven't looked closely enough at the features and attributes of this widely varied assemblage to figure out which species are worthy of respect and which are not. The tendency, in fact, has been to make the natural but unfair assumption that the proper division is between humans and all other species. Humans are creatures of moral status—they count in moral calculations—while all others are morally irrelevant, mere objects to fill human needs. Philosophers describe this view as "anthropocentric" (that is, human-centered). Those who abide by it—nearly all of us—are guilty of a form of discrimination termed "speciesism," at least according to critics of this view. We assume that our species is different and better than the others, so much so that we count and they, pure and simple, do not.

One solution to our discontinuity with nature, according to this method of thinking, is to redraw the line. We'll still have two categories, but humans will no longer stand alone in the group that counts for something morally. Humans will be joined by other species—by large apes, by whales and dolphins, perhaps by dogs and cats, and by who knows what else. Somewhere the line will be drawn.

Before talking about exact lines, however, we need to back up and see how and why philosophers have reached this intellectual point. The moral philosophers who engage in this reasoning by and large begin with the individual human, to whom they attribute moral worth. They then progress outward in ever-widening circles to see how far they should logically proceed in attributing similar moral worth to other animals. Their aim is to identify the particular factor in the makeup of humans that makes them so special, the single, most defensible criterion that captures the essence of moral worthiness. Having found this trait, they can then decide whether other animals share it. Those that do will enter the golden, favored circle that brings moral entitlements; those that do not, remain pawns. Because they extend moral worth beyond humans, these philosophers are sometimes referred to as "extensionists."

We should be quick to note that many modern ethicists believe that humans are *sui generis*—unique—and that moral worth does not extend beyond humans. We can refer to these philosophers as the exclusivists. Exclusivists sometimes have difficulties with humans who are unconscious and incurably ill and quite often dispute the status of the fetus. (None takes the position, at least so far, that humans count only if they actually live up to the species name and be sapient—that is, wise!) However they phrase the distinction, exclusivists perceive something about humans and their special mental abilities that sets them apart. Ethicists who draw the line with humans typically perceive the natural world as possessing value only as an instrument for humans to use in pursuing happiness. Any limits on our destruction of the planet stem from ethical duties we owe to other people—or in religious schemes, from duties we owe to God.

Environmental philosophy begins as a distinct discourse once we transcend the thin, arguably indefensible line separating humans from the rest of nature. Some philosophers who take this step contend that the key attribute of humans is subjectivity, our capacity to be the *subjects* of our lives. They are prepared to extend the moral worth of humans to those higher mammals that exhibit this mental trait. There will be trouble, of course, identifying exactly which species are self-aware subjects of their lives, but the problem does not seem intractable. Other philosophers extend the circle wider by identifying sentience or the ability to feel pain as the key trait. This brings in a far wider range of animals, although it still leaves a fair-sized group that is worthless.

The expansion of the elite circle to encompass all organisms that experience suffering is by no means the inevitable end of the process. In truth, many other grounds for division are possible. We can suggest the breadth of options by noting that for some philosophers the issue is not so much the functioning of an organism's nervous system as it is the organism's ability to possess goals or to exhibit distinct interests that humans can either hinder or foster. With this reasoning, we can reach out not just to all or virtually all animals but to plants as well. After all, plants seek to grow and prosper as much as the rest of us.

Although extensionist philosophers disagree on the key trait and on how far to expand the favored circle beyond humans, they all recognize moral worth in a group of nonhuman species, be it small or large. In doing so, they conclude that humans have obligations to respect these species, although not necessarily to treat them exactly as humans. Respect for a few species is not the same as respect for the Earth, to be sure, and more than a few animal welfare advocates seem little interested in life beyond their cluster of favored mammals. But depending on how we articulate the moral entitlements of these worthy species, we can end up with rigorous restraints on human conduct. For example, if we include in our favored circle the northern spotted owl that has caused such furor in the old-growth forests of the Pacific Northwest, we might easily conclude that the moral claim of the owl entitles it to an undisturbed habitat—which means, among other things, no more habitat-altering timber harvesting.

We might term this philosophic approach the "insider-and-outsider" or "two-group" approach. We can capture its essence by envisioning our imaginary mass assembly of all species as divided into two groups: one that counts, and one that does not. Wherever the line is drawn, we end up with insiders and outsiders, us and them.

This two-tier image offers a couple of vital lessons. The process of examining ourselves as humans and determining why and whether we are special helps us to understand the close ties that link us to the rest of the natural world. The more we study this issue in a serious way, the more we spot parallels and shared traits among species. The more we study, the more we see that humans are less unusual than we thought and that life in all its forms is special, even if some forms strike us as a bit more special than others.

If our natural impulse is to treat the planet as a collection of objects to exploit, ignoring the seemingly useless parts, the two–group image challenges us to rethink. The line between humans and others is indeed a slim one; the task of drawing and justifying the line is surprisingly difficult and sobering. Even if we end up leaving the line where we found it, between humans and all else, we can do so only with the uneasy sense that the line means far less than we once thought.

One problem with this two–tier image stems from the obvious line–drawing enigma. The further away from humans we get before drawing the line, the more varied are the animals we include in the favored category and the less satisfying and convincing is the line of reasoning needed to claim that they all share moral worth. In shifting outward from humans to include more and more species in the favored category, we find at each step divisive arguments about whether the similarity among included species is too weak and whether we have simply gone too far.

Even if we cannot establish a satisfactory line between the species, this two–group analysis can still serve a purpose by getting us to question our long–standing assumption that humans are the only creatures who count. There is a second aspect of the two–group image, however, that is more troubling: Once a line is drawn between the species, we must face up to its implications. Just as the homes and fields of humans cannot be bulldozed without their consent, so too other species included in the magic circle should be entitled to undisrupted habitats in which to live and feed. If the favored group of species is large and if all the species are on an equal plane with humans, this no–disturbance rule could hamper humans considerably. The more species we respect, particularly if we actually grant them rights, the less freedom we will have. If wolves are as worthy as we, if their suffering is as important as ours, how can we harm them?

We face, in short, a difficult trap. If we limit the favored nonhuman species to the largest apes, the conflicts will prove manageable (particularly, of course, in countries that have none outside of zoos). But only if the favored group includes a large number of species, widely dispersed among the world's inhabitants, can our enlarged moral sense really lead to improvements in our dealings with nature as a whole.

Wherever we draw the line, there remains a final risk in the two-group conception of the natural order: however big the favored class, the two-tier image still leaves many species without moral worth, and this group probably will be large. It is hard to imagine a healthy natural order on the planet when some forms of life are treated no better than rocks. We can save the whales by recognizing their moral worth, but we must be cautious in doing so that we do not condemn by omission all other species, and perhaps the planet's health as well.

Sheathing Ockham's Razor

The two-group debate is particularly intense because the proponents of the approach share a vision of a special type of grail. What they seek, as we've seen, is a single principle that distinguishes between things with and without moral worth. Holders of moral worth enjoy high status; things of no worth are ignored. In the value scheme of these single-line philosophers, the drop-off can be immense.

It is easy to see why arguments among these philosophers are so divisive. Animal rights advocates, mixed up in the middle of this debate, are easy targets of ridicule when they claim to perceive some morally meaningful difference between a chimpanzee and a dog or between a deer and a rabbit. The gradation among species is slight if not arbitrary, and it is disturbing for such grave consequences to rest on so little.

Some scholars—a growing number, it appears—have drifted away from the single-line camp because the whole process seems inevitably unsettling. Their dissent is not so much from any particular line as it is from the idea that we are limited to only one. The alternative approach is some form of gradualism, a philosophic scheme that places organisms in multiple categories and that recognizes gradually declining moral value as we move further and further from the human core. Instead of two groups, we now are offered many, with the rights and entitlements declining in strength as we move down the line to the increasingly less favored.

The single-line quest derives in part from the dominant Judeo-Christian heritage in the West, which drew a sharp moral line between humans and the rest of nature, thereby creating two categories. Humans had a soul; other organisms did not. As the idea of the soul withered in the face of science, philosophers began searching for another distinguishing factor that set humans apart, thereby retaining the two groups.

The idea of dividing species into two categories also draws upon the thinking of a medieval theologian, William of Ockham, who articulated what became a fundamental tenet of scientific inquiry. Ockham advised that we should prefer the simple explanation of nature over the more complex one when both seem to work. Nature, he believed, operated under simple principles; the more simple and economical the explanation, the more likely it was to be true. This rule of preference is known as Ockham's Razor, and it has proven its worth in many scientific settings. In our case, the simplest rule is the one that offers a single line between the moral haves and the moral have-nots.

Ethicists have tradtionally favored a single-line system for more than just its economy and elegance. A gradualist system creates multiple moral classes, which means more lines to draw and whole new categories of rules. The more lines and rules there are, the more complex things become and the more prone we are in particular settings to fabricate a new category or rule or to pick and choose among existing options in order to find a justification for what we want to do. The danger is relativity—the lack of rules that really bind—and multitiered schemes commonly must defend against this charge. In the single-line camp there are things that count and things that do not, and the clash between the two produces a clear winner. In gradualism we have harder issues. How do we balance, for example, a moderate need of a higher species against a stronger need of a lower one?

Humans like all animals need a steady flow of protein, which they get only from other life forms, animals or plants. But eating can be more or less efficient and honorable. With plenty to eat, perhaps it is morally wrong for humans to prey on endangered species, and perhaps it is wrong as well to eat the sturgeon's eggs while wasting the flesh. Given that humans can thrive without meat—and are likely to live longer without it—perhaps it is even morally wrong to kill any animal for food.*

In the case of human needs less vital than food, the questions become even harder. How much can humans kill and disrupt in the quest for warmth, comfort, ease, and entertainment? Can we exterminate all wolves to eliminate the modest risk that a human child one day will be killed, or is this as sensible and natural as trying to avoid falling tree limbs by cutting down all trees everywhere? Can we eliminate large habitats, thereby killing indirectly, simply to build more vacation spots?

In a gradualist, multigroup scheme of value, these questions are hard to answer. And in answering there is considerable danger of seeming if not actual rationalizations. If a moral scheme becomes too fragmented, the tendency is to develop new rules for each new situation—a process not likely to produce rules that are fair and coherent.

To assess the rightness of gradualism, we need to mix our deliberate thoughts about moral value with a few simple

---

* Not incidentally, the environmental costs of meat production are high, and the lower we eat on the food pyramid, the more gentle our effects on the Earth. According to one study, one pound of steak from steers raised in feedlots costs five pounds of grain, twenty-five hundred gallons of water, the energy equivalent of a gallon of gasoline, about thirty-five pounds of topsoil, and an unknown amount of pollution from farm chemicals.

intuitions to see where we end up. Although our senses do not organize the world into neat packages, they do send us many messages. On an intuitive level, the gradualist, multi-group approach seems to offer a good deal of accuracy and truth. That intuitive sense supports more logical claims that, as levels of functioning decline among less-developed species, moral value should decline as well.

What we are likely to sense intuitively when we survey the Earth's many life forms is a large element of inequality among the species. Bears and elk, it seems, are somehow more precious than mice and beetles, even if we do not know exactly why. The towering oak is more valuable than the patch of common dandelions. A wild stream, we sense, is valuable for what it is, and its value increases when other wild streams are blocked, diverted, and polluted. An individual condor is more valuable than its bald eagle kin, again because of its relative scarcity.

When we rely on the senses we are likely to come up with many categories, if not a gradual sliding scale. A chimpanzee has value, but we are unlikely to place it exactly on par with a human. The gray squirrel lives in a lower notch still, in part because it is ubiquitous, with most snakes and insects following somewhere below. An animal's value can depend on the size of its family. When a mule deer population becomes excessive and its habitat degrades, something must and will change. Mountain lions and wolves could keep the deer population in check, but if they are gone and the deer have few predators, humans somehow must step in to bring about a reduction to avert mass starvation. Value lies in a healthy deer population that keeps within natural fluctuations, and individual deer may need to give way for the health of the species and the land.

By inclination, in short, we approach the world from a gradualist, fragmented perspective. Our categories may be vague and fluid; we may rarely give them thought. But few of us would intuitively develop a hard-and-fast line, even a line that distinguishes humans from the rest. Only the most hardened, devoted follower of Ockham is likely to claim confidently that an animal's suffering is morally irrelevant except as it affects humans; that the screams of an animal in pain (as Descartes asserted) are no different from the squeaky gears of machinery in motion.

If we undertook to divide all species into groups, the disagreements, needless to say, would be spirited. Imagine the disputes about which species seem more "human" or just plain more warm and lovable. In addition, we would need to anticipate changes in our evaluations over time as we came to perceive that certain species were more useful, more aesthetically pleasing, or otherwise just better than we had once sensed. As we increased our feelings of kinship, our values might become more sympathetic, leading to fewer, more inclusive categories.

Part of the qualms people have with the animal rights claims—and the reason why many animal welfare groups deliberately avoid the term "animal rights"—is that most people feel uneasy with the notion that nonhumans can have human-type rights. We sense that animals count for something, but do they really have rights in something like the human sense? If we recognize rights in animals, will we not inevitably diminish the importance of rights among people? The multigroup image provides a solution to this problem, however messy and awkward it may prove in application. Animals can have moral worth, "moral considerability" as philosophers often term it, without holding full rights. We are far more likely to recognize value in a rat or even a dandelion if we can place it in a value category far beneath us.

The multigroup conception of nature can be established so that all parts of nature have at least some moral value, even if far less than humans. This sense of value in all of nature is helpful and likely to form a central part of a mature, usable environmental ethic. Standing alone, though, the multigroup image provides only modest guidance, for it doesn't tell us how to set up the categories and how much value to give to the species at each level. Nonetheless, when matched with other ideas, with visions of land health and sustainability, it can help us develop solid rules for right and wrong living.

Species and Ecosystems

The single-line and multigroup conceptions both place moral value at the individual level, with the specific dolphin or condor or Osage orange. Although an individual's value depends to some extent on its species—and, as we've seen, different species themselves can have widely varying values—ultimately, value in both conceptions resides in each individual separately, not in the species of which it is a part.

For many centuries, Western philosophy has assumed that moral worth in the universe lies at the level of the individual organism. This assumption became stronger with the advent of modern science and its claim that all matter was composed of atoms and molecules, the building blocks of life. Modern science promoted a focus that was atomistic and individualistic; both human society and nature as a whole, it seemed, were mere collections of distinct, discrete objects interacting according to the physical laws of motion, however we might categorize them scientifically.

This individualistic focus began to weaken as evolution studies showed the connections among species. But it has been the new science of ecology, born in the twentieth century, that has been the primary stimulus for a new view of life. Ecology is the study of how an organism interacts with its environment. The principal lesson of ecology is that no organism stands alone, no organism can live for long without constant interactions with other organisms and inanimate life, to gain oxygen, water, food, and shelter and in due course to reproduce. Many species live in symbiotic or social relationships and would die in isolation. As more is learned about the complex interactions among life forms, it becomes increasingly clear that the line between the individual and the environment is thin and indistinct. Indeed, for some scientists an individual organism is less a separate piece of life than a focal point or node through which nutrients and energy flow; the individual is fully defined by its interactions.

In important ways, the building blocks of nature are species, communities, and ecosystems. Many philosophers have picked up on this point to assert that moral worth really lies at these higher levels of organization. It is not the individual deer that counts; it is the deer herd or the fir–spruce–aspen ecosystem of which it is a part. Individual animals come and go; value and duty lie in the maintenance of a healthy population in a livable environment.

Philosophers who embrace this view are termed "holistic thinkers." Some holistic philosophers believe that all value rests at the species or ecosystem level and that individual organisms get their value only indirectly, by being a part of that larger entity. Others are prone to divide up moral value at the first stage, giving some to entities and some to individual organisms, with some organisms perhaps receiving more than others (as in the multigroup conception) based on higher levels of mental functioning. What distinguishes all of them is a desire to protect healthy wild populations, which translates readily into a concern for protecting natural areas and habitats rather than individual specimens.

Within academic circles this push to attribute moral value to species and ecosystems has not gone unchallenged. The problem, in brief, is that species and ecosystems are intangibles; they are mental constructs or categories that have no physical existence apart from their individual components. A black-footed ferret is a physical object, with a functioning nervous system and an ability to feel pain. The species of black-footed ferrets is nothing more than an idea, an intellectual way to organize all the individual, real-life members. If moral value arises from an ability to think or to feel pain, no species or ecosystem has value.

Holistic philosophers offer varied responses to this chal-

lenge. Some claim that species and ecosystems do have interests and preferences that are sufficiently strong and distinct
to form the locus of moral value. Others attribute moral
value in a far different manner, by relying on their deep-
seated sentiments about the functioning and interrelatedness
of all life on Earth, sentiments that have been distinctly
informed and molded by the lessons of ecology. Finally, there
are some who claim that species and ecosystems have intrinsic moral value because ecologically informed human
observers are around to create and recognize that value.

Ultimately, the details of these challenges and refutations
are of only modest importance. Whether we learn the lesson
from philosophy or obtain it more directly from ecology, it is
clear today that the welfare of each individual organism is
linked to the health of its surrounding habitat. In the long
run, organisms of all types will thrive only if the basic conditions for life remain clean and ample, only if there are nutrients, water, and good places to grow and reproduce. Given
the links among organisms and the constant interactions
between an organism and its environment, it is hardly possible to conceive of the health of a few organisms distinct from
the health of the whole. Even hard–line exclusivists can recognize and act upon this practical truth.

The importance of holistic thinking, and the extent of its
break from our inherited philosophic tradition, can be seen
in the distinct strain that exists today between holistic
philosophers and animal welfare advocates. The outside
observer might assume not just kinship but shared living
quarters between these seemingly close compatriots. In fact,
their differences are considerable and harsh words have been
spoken. Many holistic philosophers, in fact, feel more comfortable with hunters than with animal liberationists.

The claim that holistic thinkers make against animal welfare advocates is that the latter have little sense of ecological
balance or are ignoring it if they have it. They don't appreciate that animals must kill one another, and populations must
be kept in balance, either naturally or, as needed, by human

intervention. If deer populations are too great, they must shrink to avoid habitat destruction, which will harm the deer in the long run. It makes little sense to feed starving deer when the starvation stems from a population too large for its home. If we want to aid deer we should do so naturally, by developing new deer habitat and then leaving them alone.

The responding charge made by those who value the individual organism is, in its starkest form, that of fascism. To the animal welfare advocate, the claim of moral worth at the group level sounds frighteningly like the political claim that the state is all and the individual is nothing. It sounds disturbingly like the idea that individual humans can be freely sacrificed in the name of the nation. If we are to sacrifice individuals for the common good, who is going to make the deadly decisions?

One obvious way around this conflict is to say that holistic reasoning only applies to our dealings with *nonhuman* nature. Among humans we can employ a different moral scheme, one that retains moral value at the individual level. If we follow this approach, as many philosophers have, one set of moral rules will apply to dealings among humans; another set will govern the dealings between humans and the rest of nature. In some manner, these two sets of rules must be knit together with what philosophers call second-order rules—rules that help resolve conflicts when ethical norms among humans point us one way and ethical obligations to nature push us in another.

In many settings, the argument between holistic thinkers and animal welfare advocates is an artificial one, easily set aside. Much of the animal welfare drive has sought to improve the treatment of domesticated animals and wild animals in human captivity. Because these animals are removed from healthy natural ecosystems, they contribute nothing to ecosystem health. Our dealings with these animals cannot be based on any duty to respect and protect functioning ecosystems; if we have duties, it is to the animals as individuals, not as members of some group. Because these animals are entirely dependent upon humans, they would seem to be in a different moral category from animals in the wild, and different moral rules would seem appropriate. On this issue, it is the reasoning of the animal welfare advocates, not that of holistic thinkers, that is likely to prove most useful.

The Magic Calculator

All of the philosophic approaches so far considered base ethical norms on the idea that moral values exist in the world and that ethical norms ought to be aimed at respecting and enhancing those values. Standing in marked contrast to these approaches is the ethical system known as "utilitarianism." In utilitarianism, ethical norms are typically established so as to enhance the overall utility or well-being of humans. An action is right if it adds to human utility; it is wrong if it detracts from that utility. Utility is measured in various, similar ways—in terms of human happiness or pleasure, the satisfaction of human preferences,. or the production of certain intellectual or aesthetic experiences.*

According to utilitarian thinking, our environmental problem has come about because we have made mistakes in calculating the best way for us to live. We have eroded our soil, exterminated species, and polluted our streams, all because we have miscalculated the net consequences—by overstating the good, understating the bad, or both.

At a superficial level utilitarian thinking has a great deal of appeal. It tells us that we need to look closely into the consequences of our conduct and make sure that the net effects are as good as they can be. We've undoubtedly calculated poorly in the past, and a closer look could well lead to better, more sensitive decisions.

To understand the value of utilitarianism, however, we need to keep aware of its rather considerable limits. One of

---

*The special views of utilitarians who take into account the pleasure or pain of nonhuman animals are included in the discussion above of extensionists.

the problems with utilitarian thinking is that it can quickly get us bogged down in very complicated calculations. In order to determine the rightness or wrongness of an action, we need first to identify its consequences. This undertaking, however, often requires a great deal of knowledge, far more in many instances than we have or are likely ever to possess. We uncovered this problem when looking at the externalities image that economists offer. We can throw an object into the swirling physical world of motion and energy and very soon lose all track of its indirect effects. We fill a wetland and put in a factory: who can enumerate all the consequences? When we plow the prairie and put in wheat, who knows all that will happen? Indeed, can I even know fully what the effects will be if I plant a birch in my front yard instead of a hickory?

At its best, utilitarian theory would function like a large, magic computer, one with sufficient calculating power to make even the most complex calculations quickly, one with an underlying data base that includes not just the entire realm of human knowledge but vast amounts that remain unknown to us as well. We do not, of course, have a computer like this, nor are we likely to develop one. And even if we did, one risk of utilitarianism would remain: this is, given its appetite for facts and figures, it could (and does) encourage us to cast off our intuitive sense of right and wrong and rely solely on long chains of facts and logic in deciding what to do. We spot a factory belching fumes skyward and must pause before we criticize. Without the magic calculator, who knows what is right or wrong? We sense trouble, but maybe our calculations are simply incomplete. We need more information and must sit still until we get it.

Part of the job of this magic computer would be to help figure out when a consequence is good and when it is not. Utilitarians do this now by figuring out whether a consequence does or does not add to aggregate human good (whether defined in terms of happiness, the satisfaction of human preferences, or in other ways). But this calculation proves challenging, even if we accept the standard of measure. When Robert Ciampitti fills his wetland, his happiness and the happiness of those who move into his new houses will go up. But there is human unhappiness as well that comes about because of the wetland loss and the ecological damage that it brings. Because it is spread widely and thinly in time and space, this unhappiness will be hard to trace and aggregate. So we are left with the question—is the transformation from wetland to houses a good consequence or a bad one?

Another important uncertainty in our calculations lies in the discount rate—the rate at which we discount future harms before we compare them with present-day gains. If we fill the wetland today, the benefits will come soon; many of the costs will be much delayed. As we noted in discussing economic images, one of the most grave problems in protecting the environment is that commonly used discount rates are all sufficiently high that the future—even as little as twenty or thirty years out—counts for almost nothing. In using the utilitarian's magic calculator we will want, in all likelihood, to program in a discount rate; the calculator itself cannot help us pick one. We must decide for ourselves how much we want to value future healthful planetary life and somehow incorporate that valuation into our calculations.

Perhaps the most important limit of utilitarianism is that, like the economists' free market, this line of thinking is morally neutral in environmental terms. Utilitarianism seeks to give people the most of what they want. If people want to live in a healthy natural environment, it can help them achieve that. If people are made happy by soiling and consuming the Earth, the magic calculator can help at that as well.

Although these limitations on utilitarianism are all important, the cost-benefit thinking inherent in utilitarianism will certainly be a part of any good ethical approach to planetary health. Before using it, however, we need a clear sense of values and goals, a clear vision or image of what we want to achieve in terms of lasting land health. Once we gain a vision of sustainable life, we can use the calculator, pro-

the next paddler who comes by—if one does come by during the few days I am here.

My spiritual roots sink into this rock ledge beside Clearwater Lake, and I soak up, slowly, a sense of caution and awe.

5

Preserving the Wilds

Possibly, in our intuitive perceptions, which may be truer than
our science and less impeded by words than our philosophies,
we realize the indivisibility of the earth—its soil, mountains,
rivers, forests, climate, plants, and animals, and respect it collectively not only as a useful servant but as a living being.
—Aldo Leopold

The future generations image is closely linked for many people with the vision of wild nature, the raw Earth unaltered
by the axe, the plow, or the concrete paver. The link between
the two images—future generations and wildness—is an easy
one, for the clearest way for one generation to pass on as
good as it received—if that is its burden—is for it to preserve
everything.

It is time at this point to look more closely at this preservation image, this vision of the pure and the pristine. We can
do so by picking up the image as it arises and is put to active
use in one of America's most remote locations.

The Burr Trail

The Burr Trail is a one-lane, sun-baked Jeep track that cuts a
wandering way through the barren plateau of southeastern

Utah. It runs through sagebrush country, the land of juniper and piñon pine, of the jackrabbit, the sage grouse, and the occasional pronghorn and mule deer.

A traveler can begin the Burr Trail at its northwest end in Boulder, Utah, a sparsely populated town in near-empty Garfield County. The trip demands a vehicle with heavy tires and a stiff suspension, for the Burr Trail's appearance on highway maps can be misleading. No bulldozer or earth grader has ever found time to iron out the many wrinkles of this naturally rugged terrain. For sixty-six slow miles, the traveler feels every rock and gully, like the corduroy roads of old. In summer the dust clogs the pores, and the distant hills shimmer in the dry heat.

At the southeast end of the trail is Bullfrog Basin Marina, situated on stark, sun-drenched Lake Powell, a reservoir created by the flooding of the former Glen Canyon. Spaces are big in Utah, and with its reach of more than two hundred miles, Lake Powell is no Walden Pond. No intermediate stops interrupt the trip between Boulder and Bullfrog, not so much as a town or crossroads. The traveler finds no signs like the one on Interstate 70, miles to the north: No Services Next 110 Miles. Out of habit, one can only surmise, the planners of Interstate 70 added exits every dozen miles or so, but many of them lead nowhere and, by all appearances, hook onto nothing. The exit signs announce a plausible purpose, "Ranch Exit," yet one searches the horizon and spots no signs of life.

On the Burr Trail, signs and exits seem like the unneeded fluff of a congested and distant world. The shoulder is like the road, the countryside is like the shoulder, and all blends together to give rise to images of freedom and unlimited options. This is Bureau of Land Management land, so dry and unproductive that even sparse numbers of cattle can (and are) overgrazing it to death.

The land surrounding the Burr Trail seems untamed, even brutal, but its power is like a desert mirage. Punch it and it stays punched. The arid country is fragile and shows its scars. In a place of plentiful rain, a Burr Trail would soon be overrun by weeds and torn by erosion. But aridity defines the West, particularly the ranges and uplifts of the Colorado Plateau. A few passes of a Jeep across the sage flats leave a path visible for years. After a few dozen vehicles, perhaps one per week, the trail becomes a landmark that might last for a century. Aside from the ravages of overgrazing, the countryside is stark and the air is shockingly bright. This is wilderness of the open rather than wooded kind. Withal, the country is beautiful or repulsive, rich or barren, valuable or valueless, all based on the judgment scheme brought to bear. Like all wilderness areas, it carries intellectual and spiritual messages in abundance.

From a distance it is hard to visualize much commotion about such an insignificant path as the Burr Trail. But the viewer from a distance is likely to think of the trail as simply a route from point A to point B. Like most distant looks, this one leaves out much. On its way to Bullfrog and Lake Powell, a major recreation area, the Burr Trail passes through Capitol Reef National Park, by two areas under study as potential wilderness reserves, and by numerous ruins of the little-remembered Anasazi occupancy.

Today, the Burr Trail is under siege. The Sierra Club sees beauty out on the sage flats, beauty that is disrupted by the presence of even a rarely used track. Garfield County anticipates restraint and frustration when it contemplates increased wilderness protection. Wilderness is a legal as well as mental construct, and wilderness in the law means an area off limits to those who seek work. Wilderness, the federal statute says, is not just land "retaining its primeval character and influence, without permanent improvements or human habitation"; it is land "where the earth and its community of life are untrammeled by man, where man himself is a visitor who does not remain." All species but one can make a home in the wilderness, which is just the way environmental groups want it. They embrace this definition with vigor: a single hut, a road or sign—anything can spoil the wilderness experience, that spiritual sense of being detached from all

humans do and represent. The Burr Trail runs by two wilderness study areas and, except for the road itself, nearly all the
surrounding land meets the statutory definition of wilderness; nearly all of it, one day, could be added to the wilderness inventory, which means strict protection in perpetuity.

Garfield County seeks to widen the trail to two lanes, to
level the bumps, and coat it with gravel. A distant viewer
might guess why this is so, but the guess again is likely to be
wrong. Travel on the trail is light, and local drivers are prepared. In fact, the trail needs no upgrading to satisfy local
users, and the county doesn't view its proposal principally as
a local convenience. To the county, the trail means more: it is
a symbol of what might be. And what might be is recreational travel and dollars as people to and from Lake Powell
spread their money. If gravel is added to the trail, the local
dream goes, people will come, and the county will be like
the rest of the country, prosperous and forward looking.
Without the improved trail, the county will be condemned if
not cut off, confined in its ability to chart an American
course of growth.

And so the lines are drawn, the spiritual quest for
human-free land and the ever-present hope for economic
growth. Vision against vision, image against image. With
both images absolute, compromise is illusive. The Burr Trail
has become like the smokestack—a symbol of jobs, industry,
and opportunity to some; of pollution, sickness, and degradation to others.

The Spirit of Wilderness

In the 1930s, Robert Marshall, Aldo Leopold, and a handful
of other environmental pioneers got together to form a new
group, the Wilderness Society, which was to work for the
preservation of wilderness areas in the United States.
Marshall was no stranger to wilderness. With independent
wealth behind him and strong legs beneath him, he set
records for hiking mountain peaks all across the land.

Leopold, too, was familiar with the idea, for he was instrumental in pushing the United States Forest Service in the 1920s to set aside the nation's first primitive area, in New Mexico's Gila National Forest. Marshall was the adventurer; Leopold the scientific forester turned wildlife manager and wilderness advocate. Had John Muir still been alive he, too, would have been among the first members, for his spirit, the spirit of the mountain man escaped from nature and living on the fringe, permeated the movement from the start.

In the decades since the 1930s, wilderness preservation has become a beacon of the environmental crusade. In 1964, Congress adopted the Wilderness Act and selected nine million acres of National Forest land to begin the new system of preserves. The Forest Service in the 1960s and the Bureau of Land Management in the 1970s reluctantly began reviewing their landholdings to identify lands that seemed to meet the statutory definition of wilderness, which could be recommended to Congress for inclusion in the system. Year by year, new pieces were added, mostly in the West and disproportionately in Alaska. By 1990, the system included nearly one hundred million acres, twice the size of Nebraska, nearly twenty times the size of Massachusetts.

Although the wilderness crusade stems from a love of raw nature, it has gained much of its strength because it offers to environmentalists such clear signs of progress. In the 1970s, environmentalists obtained new laws to clean the air, clean the water, and moderate the flow of hazardous wastes. As the years went by, however, progress was slow— halting and at times even backwards. Results were illusive and questionable, except in the case of a few waters like Lake Erie that could do nothing but get better. Even less success was achieved in halting urban sprawl, in saving prime farmland, and in reducing soil erosion and groundwater exhaustion and pollution. The tasks were enormous, the fights were endless, and the signs of success appeared contradictory.

The wilderness quest, in contrast, brought identifiable, solid gains. Each acre added was a victory; once placed in the win column, it was there more or less permanently. Roads, buildings, and waste dumps were out; even huts and signs were kept at the gate. As the list grew longer and the total acreage increased, the sense of pride and progress strengthened. Here, it seemed, was unambiguous success, something that couldn't be denied.

On the private front, a miniwilderness project was being undertaken by a thriving nonprofit group, the Nature Conservancy. This group engaged, not in lobbying, litigation, and confrontation, but in land acquisition. It preserved the land in the most direct of ways: by buying it. As frustration followed frustration on the pollution front, the conservancy's membership swelled.

The cause of wilderness preservation attracts the ardent environmentalist but also, surprisingly, those who worry that stricter pollution laws will mean job losses and industrial decline. The appeal is wide, partly because the wilds can be preserved without disrupting our daily lives. Wilderness restricts the future, not the present, and it allows us to relish our altruism without giving up a thing. Wilderness may seem like the most extreme form of environmental commitment, but to the average American it is the cheapest.

For people content with what they have and with the choices they can make, the sacrifice involved in wilderness preservation can be trivial. The sacrifice comes only from the comparatively few people who, out of work and scrambling for options, look to the wilderness as the near–term chance. And here arise some of the most symbol–laden and, hence, acrimonious debates on the environmental front today. Miners, loggers, ranchers, recreational developers, and others who sense money to make typically offer harsh words about the silliness of wilderness. On the opposing front, wilderness proponents talk of greed and pride, of the inevitability that the gold of wilderness, once touched, will turn to stone.

Amid the pushing and posturing, it is easy to overlook in this debate the shared idea of what wilderness is and what it is not. Wilderness to both sides means nature unaltered by the human touch. Wilderness is not the pastoral ideal of nineteenth-century poets and painters, the rural village undefiled by the ravages of industrialism. Neither is it a countryside occupied harmoniously by an isolated Native American clan following age-old practices. Wilderness is nature with no signs of the inevitably dirty human; it is nature with the people erased.

Whatever the values of wilderness—and they are considerable, as we shall see—the preservation image is incomplete and hazardous, which means that it must be handled with care. People do not live in this type of pure wilderness, where all signs of human alteration are viewed as defilement. Because humans are absent, the image doesn't show us how to live. As importantly, the successes in preserving wilderness can easily lull us into thinking that victories can come easy, by the flex of political muscle and the stroke of a legislative pen or by the dispatch of a modest check. When the real work begins, the strains will be greater than this. And so will be, we can hope, the rewards.

The Benefits, Near and Far

In a scientific vein, wilderness offers a bank of unexplored and unvalued genetic material. Its organisms, whether named or not, may some day help heal our ailments, fill our stomachs, or clothe our backs. A wilderness area is a scientific bonanza, a stack of Rosetta stones awaiting translation. It is a newly discovered library yielding its secrets, slowly and coyly, as scientists take time to look and listen. Wildlife from the wilderness adds to the richness and diversity of our larger ecosystems and helps buffer human excesses on neighboring lands.

Wilderness shows us species, large and small, living within their means. It shows animals that have developed ways to sustain life without ruining their environments. Some of these lessons can be directly applied. The buffalo developed ways to graze Western lands without ruining the range;

many of our cattlemen and their cattle have not. Other lessons are more subtle and need to be generalized. For example, at times we see species that explode in population and overuse their habitats, which leads to starvation or other means of rapid decline. In this, too, there is a chance to learn.

But if wilderness offered only utilitarian benefits, it could hardly have captured the human imagination as it has. Its primary benefits are more emotional, spiritual, and religious. In a seemingly infinite variety of ways, wilderness can move the visitor in the innermost part of the soul. The wilds bring solace and inspiration, hope and stability, a sense of wonder and a sense of place, a spirit of contentment and a spirit of adventure. We can sense God and see the power of Creation; we can see ourselves, perhaps more clearly than ever, and wonder at the complexity and the promise. We can reach out to other species and feel like tiny specks in the vast, rich array. By opening ourselves fully, we can sense these things and more, and the feeling can linger, nature's healing and wholeness at work.

Wilderness inspires the pen as well as the heart, and to this day the flow of wilderness memoirs is brisk. Through this literature, the reader can readily gain a sense of the many ways that people find meaning and solace by touching the raw. Even the mere knowledge of distant wilderness can inspire people, as evidenced by the many wilderness advocates who have never set foot in one and may never do so. "We simply need that wild country available," wrote Wallace Stegner, "even if we never do more than drive to its edge and look in. For it can be a means of reassuring ourselves of our sanity as creatures, a part of the geography of hope."

Just knowing the wilderness is out there somewhere can help us enjoy the Earth, an enjoyment made more vivid by nature films. I can experience loss at the poaching of the black rhinoceros, whether or not I harbor hopes of an African visit. Added to this is the simple beauty of wilderness, in its vistas wide and narrow. Wilderness offers beauty also for the ear, in the call of the loon and the coyote, in the soothing, pulsing rush of a hurried mountain stream.

Part of the value of wilderness is that it offers evidence that there are actually limits on our greed. When we preserve the wilderness we have at least placed a modest rein on our consumptive appetites. We have spared something in what has seemed, until recent times, an almost uncontrolled quest to exploit the entire landscape. This particular wilderness value—wilderness as evidence of character—is enhanced by the fact that wilderness preserves are publicly owned. Indeed, the value might largely disappear if we decided to turn wilderness areas over to private hands, for the restraint then would be private rather than communal. Americans can be justly proud of Yellowstone Park because it was the first national park in the world; our country originated the idea, one that showed foresight, sensitivity, and restraint. These are virtues that we admire and would like to embrace, and it makes us stand tall as a people when we can point to instances where we have actually displayed them.

With these benefits, wilderness preservation can make sense even to the calculator–toting, anthropocentric utilitarian. More intuitive, emotive observers require even less persuading: they soak up directly the beauty and rightness of it all. Mere exposure, and they believe.

Wilderness and Right Living

Wilderness preservation, to be sure, will be a part of our harmony with the natural world. It will be one of our authors, a vital, irreplaceable chapter in our textbook. But wilderness will not be all of the text.

One of the implications of the wilderness image is that the presence of humans in the landscape is a bad thing. Wilderness means freedom from all signs and symbols of human occupancy. Airplanes hundreds of feet above are not permitted, even though their presence leaves surface wildlife and ecosystems unaffected. The Burr Trail, however modest and little used, is enough to spoil Utah's sagebrush plateau.

We might well ask why we are so insistent about the purity of wilderness preserves. Properly constructed, the Burr Trail will not materially alter plant life on the plateau, nor will it cause noticeable erosion or air or water pollution. Few animals will be bothered by its presence, for the pronghorn, the mule deer, and the coyote all can live with modest human disturbances. Some animals will not nest near the trail, but they will still have miles and miles in each direction in which to wander and hide. The trail's effects on nonhuman nature would be almost trivial.

In light of the slightness of this disruption the environmentalists' real concern, plainly, lies elsewhere, with the human values that wilderness fosters. The spiritual values that arise inside us, within our world-toughened hides, can be as fragile and sensitive as the finest glass. To the canoeist paddling in northern Minnesota's wilderness canoe country, the slightest purr of a motorboat can dispel the illusion of turning back the clock to the era of the eighteenth-century voyageur. The sounds of modernity can too easily disrupt the sense of oneness with nature. Because of the special spiritual magic that only full wilderness can create, we want our wilderness areas kept pure. We want wilderness as an escape, as a diversion, as a monastery to which we can retreat. We want wilderness to be a nonhuman counterpoise to the human sphere, and its success in this role, many agree, depends precisely on the complete absence of all human signs. A healthy, natural ecosystem simply won't suffice. Indeed, strict limits may be needed even on the number of backpack-carrying visitors.

Once we understand this spiritual need for purity, it is easy to see why wilderness preservation is a useful but incomplete guide. Pure wilderness presupposes the existence of nonwilderness; it draws its identity and definition from comparison with a nonwilderness area that humans inhabit, whether that area is a stable, harmonious natural order or one that suffers from degradation. Wilderness and non-wilderness are distinct ideas and distinct places.

What the wilderness image does not offer, what we need to add in some manner, is the all-important distinction between use and abuse by humans. In the wilderness image, all that humans do is abuse. And in the end, as our guide, this simply won't work. We need places on Earth to live and enjoy, which means that we need a workable, sustainable vision of using the land.

A more practical problem with wilderness preservation is that it might furnish us with an excuse to misuse nonwilderness lands. We might decide that, having paid our dues to the planet by preserving bits of the wild, we have earned the right to alter other lands as we wish. If this happens, we may end up in the unfortunate condition of having two extremes of land—the fully preserved and the abused—when what we need most of all is land in the middle, land that we use in ways that are lasting and healthy.

Gaia and Deep Ecology

Related to the idea of wilderness preservation is one of the most intriguing visions put forth by environmentalists, the Gaia theory, named after the ancient Greek goddess of the Earth. For millennia, scientists and their early progenitors have pondered the possibility that the Earth is simply one big organism, with parts and systems that relate and interact in some sort of orderly or programmed manner. Perhaps the Earth is not a huge, inert hot rock, inhabited by randomly developed species. Perhaps it is actually an organism itself, Gaia, and the plants and animals are all indispensable components that interact and regulate the whole.

If the Gaia thesis at first seems absurd, we might consider that the human body itself is composed of billions of parts. Each cell includes uncountable molecules. Cells form tissues, tissues form organs, and organs form systems and ultimately the whole itself. If we could attribute intelligence to a cell in the midst of a kidney, what chance is there that the cell could grasp the organization of the whole?

The recent popularity of the Gaia thesis comes largely from surprising evidence about the evolution of Earth's atmosphere. Over the eons, the composition of the atmosphere has changed slowly, from one free of oxygen to one with almost precisely enough oxygen to sustain large–scale, active animals like humans. Much more oxygen, and fires would burn without stopping; much less, and active animals of the sorts we know could not live. Despite cataclysmic events on the planet and seemingly long odds, the Earth has remained a healthy place for life.

This atmospheric evidence, along with other intriguing if ultimately inexplicable data, suggests to a few scientists that there is more going on here than good fortune. Perhaps the Earth has self–regulating mechanisms of which we are unaware, mechanisms that somehow slip between the pictures in our microscopes and those in our telescopes.

For our purposes, the evidence and even the truth of the Gaia idea are less important than the idea's value as a stimulant and guide. The Gaia image has surfaced and resurfaced throughout recorded human history. It carries an intuitive and emotional appeal that transcends the known facts; it is one of the ideas that help us deal with the great unknown. In this secular, scientific age, the unknown has become a foundling to that proud child that we call truth and human knowledge. But it remains out there, lurking beyond the door. If we cannot prove the Gaia thesis, neither can we disprove it, and we're left in that common state of ignorance and wonder.

The Gaia image sends us contradictory signals. If the Earth is a random collection of rocks and organisms, perhaps we can rearrange the rocks, tame and eliminate many organisms, and end up with a more enjoyable place to live. If the Earth is, however, an organism itself, perhaps we invite calamity by playing the tinker and rearranging the parts. Perhaps one of the components we destroy will prove to be a vital, nervelike element that keeps the place hospitable. With this possibility in mind, we are encouraged to tread gently and to assume at least potential value in all we see. Perhaps, just perhaps, the northern spotted owl is a node for essential, global forces that are undetectable by our feeble senses. If humans too are part of Gaia, perhaps there is a prescribed role for them, a role that they must play to avoid systemic disruption. As chief proponent James Lovelock notes, "Gaia theory forces a planetary perspective."

Yet, like Janus, the ancient Roman custodian of the universe, Gaia offers us a second, oppositely directed face when we turn it around. If for eons the Earth's atmosphere has maintained its equilibrium against eruptions from the planetary core and assaults from outer space, if it is so tough and so reactive, then—who knows?—perhaps it will have no trouble coping with our determined efforts to pierce the ozone layer and warm the planet. Perhaps imperceptible regulatory forces will kick in and make whatever adjustments are needed to compensate for what we do. If this is the case, perhaps we should just continue our consumptive party and forget about the mess.

Older, more intuitive versions of the Gaia thesis are less contradictory in their messages, and people who sense an essential unity in all creation are not likely to be misdirected by the second of Gaia's faces. Many western religious figures have stressed the unity of creation through its common source in God's handiwork. All parts of the Earth, they tell us, share in God's splendor, and we disserve God by misusing what He has given. Albert Schweitzer lived a more fragmented version of this ethic by holding in his soul a reverence for all life. In many, varied ways, early polytheistic views also vested nature with inherent spiritual value.

By accepting the Gaia thesis, we circumvent the philosophic difficulty of attributing moral worth above the level of the individual organism, for it says that the Earth, too, may be an organism, the largest and grandest of all. If it is, it surely deserves the largest portion of our respect. If, on the other hand, the Earth as a whole lacks value, how can a tiny human be worthy?

Closely linked to the Gaia thesis is the image of the world put forth by the radical environmental framework known as "deep ecology." Many ideas fit under this large umbrella. Perhaps the most central is that humans are just one of millions of species, no higher or lower, and all are mixed together in an inseparable web of life. The coyote and the rabbit are equally important and every bit as entitled as the human to land, freedom, and opportunity. Deep ecology seeks to undermine human haughtiness and pride and to replace it with humans living simply and harmoniously with the Earth. For most adherents this implies vegetarianism, at least so long as the human population remains as large as it is.

Like the Gaia image, deep ecology depicts humans as one of many, many parts of a coherent, interrelated whole. All parts are important to the healthy functioning of the Earth. Humans should blend into the natural landscape and be no more conspicuous than other species—if not even less so, given that other carnivores will be allowed to keep eating meat. We must purge ourselves of our anthropocentrism and cease thinking of ourselves in elevated terms. The medieval great chain of being, in which humans formed the highest earthly link just beneath the angels, is turned sideways by deep ecology into a horizontal chain. Humans are linked to the other species, and none is higher or lower than the others. If the Earth is to remain (or become) a healthy place, humankind must show full respect for all other species.

After providing us with this helpful, inspirational idea, deep ecology begins to shift out of focus. What does it mean to say that humans are no better or worse, no higher or lower, than other species? What does it means, in practical terms, to say that we must rid ourselves of our anthropocentrism?

As we saw in our consideration of philosophic images, we encounter problems when we undertake to attribute moral worth to nonhuman species, much less when we decide to treat them as equals. So long as our concern is with only a few large mammal species, or only with captive and domesticated animals, we can respect their equal moral worth and still live comfortable lives. But when we extend this high level of moral value to a wide range of species, perhaps even to all species, we become perplexed if not immobilized. How could we possibly treat all other species the same as humans?

The tendency, when we expand moral considerability widely, is to dilute what it means to be morally worthy. The more we dilute it, the less we are restrained and guided by the moral order that emerges. Another possibility, of course, is to return to the multigroup image and grant moral considerability to other species in a series of descending steps. This approach, as we saw, accords rather readily with our intuitions and gives us the kind of flexibility we might like. But different levels mean different treatment, and different treatment is precisely the idea that deep ecology seeks to banish.

Deep ecology remains in its infancy. It may be much too soon to try to capture it all with this single image of the horizontal chain of species. Another fitting image (if we may call it that) may simply be that of the invisible human. In a sense, what deep ecology urges is for humans to blend into their surroundings so completely that they do not obtrude.

In a wilderness area, all life seems harmonious and fitting. All things interlock, with no disruptions or discontinuities in the natural order. This same feeling of harmony and proportion can arise when we watch a primitive clan or tribal group as it follows ancient customs on the land. It can arise as well when we consider a preindustrial pastoral scene, with farmers tending their fields and flocks, leading simple lives in tempos set by the sun and moon and the yearly seasons. In some scenes, humans seem to blend so naturally into the landscape as to be part of it from the beginning.

But it is only at first glance that the people in these visions seem to transcend their anthropocentrism (if they ever really had it) and treat other species as if they were human. A tribe might well refrain from overkilling the beaver because the beaver embodies, for the tribe, important spirits that would be offended by overuse. Yet they do kill some beaver in order to survive, and they probably exult in their prowess in doing so. The pastoral farmer, too, dominates other species in utilitarian fashion. The flocks in the field, perhaps a sheepdog as well, all are forced to work for human welfare. The village area is off limits to large predators, and rocks, barking dogs, and gunfire might greet any predator that tries to trespass.

The two–group image asks us to invite certain other species into the elite human fold. The multigroup image asks us to accord lesser, but nonetheless important, moral values to various categories of species. The horizontal chain image from deep ecology asks us in no–nonsense fashion to reject completely our anthropocentrism. Where the Golden Rule admonishes us to treat each other as we want to be treated ourselves, deep ecology tells us to treat other species as we do each other.

Is it in fact possible to treat other species as ourselves? Is it even desirable? Whether we can and should abandon anthropocentrism will be a central issue in the decades to come. It is an issue that, in our search for planetary health, we'll need to address head–on.

Transcending Anthropocentrism

At moments when we are tormented by thoughts of our abuse of nature, the idea of transcending our human–centeredness can strike us as the beacon from a distant lighthouse strikes the storm–tossed sailor. As we pitch and roll in a sea of confusion, wondering what will happen next, the distant light shines bright. It seems so reassuring and inviting that we cannot resist heading for it. We don't know, we don't look to see, what sharp rocks lie between us and the light.

As we consider transcending anthropocentrism, the obvious first step is to figure out what this means. At the most general level, the idea can simply mean overcoming our pride, the deadliest of ancient sins, in all its many manifestations. This idea, needless to say, is a good one, even aside from its environmental implications. We'd be far better off if we injected more humility into our dealings with individual people, with groups, and with nations, as well as with other species. A similar version of transcendence is that we should place God in the center, not humans, and that we should do not as we want but as God instructs. Both of these ideas can be good departure points, but we need to hear much more about the road ahead.

A troubling possibility is that we'll decide to tinker with our anthropocentrism simply by treating unfamiliar humans with less respect. Instead of placing humans at the center, we might just insert ourselves and our family and friends. Misanthropy is certainly not what deep ecologists have in mind, but outside observers might rightly wonder what would really come to pass if we followed their precepts. When we expand greatly the favored class, we can hardly help but reduce what it means to be favored. Will we sacrifice ourselves, push our family and friends to sacrifice, or will we expect others to bear the full burden? It goes without saying that this approach would prove unhelpful ecologically, to say nothing of being disastrous in interpersonal terms.

There are many ways that we could partially transcend our human–centeredness. We do this to an extent whenever we recognize moral worth in nonhuman nature, either at the individual-organism level or at the species or ecosystem level. As we've noted, there can be great value in this kind of limited transcendence. But going beyond that raises troubling issues.

In its fullest sense, transcending anthropocentrism means treating all other animals just as we would treat our family members, friends, and all other humans—that is, giving other animals not just some moral value but equal value. We would no more shoot a snake than a person, no more deprive a dog of life–saving medical care than a human. This extreme form of transcendence is likely to make us very uncomfortable very fast. It is hard to imagine that we could ever embrace it. The effort would require an extraordinary act of self–effacing, self–denying will—something that seems almost, if not literally, extraterrestrial.

If we decide to try to live as do other species in terms of mental or spiritual framework, this full form of transcendence seems excessive. As far as we know, no other species engages in species transcendence. When a wolf leaves its den in the morning for a day on the prowl, it sees the world through wolf eyes. To the cottontail rabbit, the world may be rabbit–centered, but this means little to the wolf, who will kill rabbits with no noticeable chagrin and show considerable partiality toward the members of its own wolf pack. For the wolf, by all appearances, it's a wolf–centered world. (Even species that eat their own kind transcend species centrism only by narrowing the favored circle to the individual organism, the I, and treating everything else as mere object.) If wolves can live in a wolf–centered world and avoid destroying their habitat, why not humans? Is there a reason why we should do more?

One answer to this forceful question is that humans are more destructive. Humans have eliminated their predators, except the microscopic kind. They have learned powerful ways to alter the environment to feed their stomachs, their needs, and their whims. Exceptional power like this inevitably destroys unless accompanied by an equally powerful form of restraint. Maybe the reason to transcend anthropocentrism is because of this special need we have for some form of voluntary, exceptional restraint.

The defect in the human make–up is not that humans look after their own interests first. The defect is that, in doing so, they fail to exercise enough restraint. Our tools are too powerful for the job, and seemingly too powerful for our moral condition: When the wolf fills its stomach, it rests or plays; its work is done, and there is no reason to do more. When humans fill their basic needs, they move on to create new needs and, having filled those, proceed to create more. After a time, the needs become so artificial that needs merge with whims, which in turn merge with extravagances.

And yet it is not primarily the proliferation of human needs that is the root cause of environmental degradation. The problem is not simply our desire for large, climate–controlled homes and convenient personal transportation. The problem comes instead when we satisfy these new needs by means of the typical rapacious striving that all species use to acquire the elements of simple survival.

At this point, we do not know what we need to do to regain a sustainable, harmonious relationship with the Earth. As we move in that direction, step by step, perhaps we'll have to cut back on many of our so–called needs. Certainly we may want to live more simply so that we can achieve some of the cultural benefits of living in a more people-oriented and less thing–oriented way.

At this early stage in the process of rebuilding ties with the Earth, it is by no means clear that those of us living in middle–class American affluence must steel ourselves for some dramatic, forced reduction in our needs. In all probability, we will be able to keep fulfilling not just our basic needs but most of our desires; we will simply need to work more ingeniously to satisfy them. Where, when, and how we satisfy the desires will certainly change, but satisfaction may still come. There are, however, some wants that will be forced out, like our desire to destroy fragile desert soils with high-powered dirt bikes.

The task that lies ahead is bigger than we can now imagine. Yet, however tentative our current vision, it appears that we need not abandon our focus on our own needs. We can continue placing humans first. What we need are not fundamentally different priorities, but fundamentally different ways of satisfying them. We need ways of living that are sustainable indefinitely, ways that promote land health rather than land sickness. The right way of satisfying our needs will be the way that promotes a healthy Earth.

In the end, it may prove easier to develop sustainable life patterns than to transcend anthropocentrism, and there is no reason why we shouldn't take on the easier chore. Our preference for humans is probably more than cultural; it may reside deep within our genetic makeup. If this is true, we might just as well try to teach the hungry wolf to turn over its food voluntarily to the even hungrier coyote. Changing our lives may prove easier than changing our characters.

Having concluded this, however, we're still left with a final use of the image of anthropocentrism transcended. The idea may be impractical, it may be excessive, yet it still can stand as a splendid goal. Many of our goals are inconceivably grand—peace on Earth, justice for all, the elimination of human evil, and the like. We can hope for universal brotherhood and sisterhood, however much we doubt our ability to get there. We strive because we must, because our souls will let us do no other.

When we set impossibly high goals, and do so based on reason rather than intuition and emotion, we set the goals as prophylactic measures. With a high target, we might end up achieving more than with a lower, more reasonable one. For many of us, transcending anthropocentrism may be just the high goal we need to summon our full energies. Albert Schweitzer held fast to a reverence for all life, and the vision led him to incredible heights. But we need to be cautious. For the average human, this goal may ask so much that it instills resignation and indeed may make no sense. In the end, perhaps we should save this ethereal vision for those special people who can make it their own.

6

Faith in Science

If we want to know and cannot help knowing, then let us learn as fully and accurately as we decently can. But let us at the same time abandon our superstitious beliefs about knowledge; that it is ever sufficient; that it can of itself solve problems; that it is intrinsically good; that it can be used objectively or disinterestedly.

—Wendell Berry

From time to time in our exploration of guiding ideas and images, we have taken shots at the vision of humankind as conqueror of nature. The time has come to examine this image more closely—to see where it came from, what it looks like in its contemporary garb, and whether despite its flaws it has strengths that appear useful. Is it really wrong, and if so how and why, to exult in our obvious, extensive prowess?

The Conqueror's Progress

In thinking about the image of humankind as conqueror, it is easy to assume that the idea is age-old, an ancient relic that has stalked us out of the distant past. When we close our eyes, into our minds may step some hulking, ill-clad fig-

113

ure. Perhaps it is a cave–dwelling hunter–gatherer, clothed in animal skins and toting a well–handled spear. Perhaps it is a different version; hoe in hand, our first ancestor who culti-vated the soil or our first ancestor to live with domesticated animals.

These early ancestors exploited the Earth to meet their needs. They ate when hungry, cut wood to build fires, and used the physical stuff around them to make life less harsh. They were earthly exploiters within the limits of their modest needs, their elementary technology, and their sparse num-bers. Compared to us, however, these Stone Age humans were infants when it came to exploitation. Our hypothetical Stone Age conqueror needed to wax and mature before he could really make use of modern exploitive methods. He needed to change, and change considerably, the way that he envisioned and interpreted the world around him.

We can understand the modern ideology of conquest, grasping all it includes, by taking a look at the various steps in the intellectual development of this Stone Age conqueror. One way to trace this growth is by constructing a brief, if somewhat ahistorical, intellectual biography.

One mental step that our budding conqueror needed to take in his march to maturity was to remove the spirits that surrounded him in the natural world. Early humans often saw deities or sparks of the divine throughout their sur-roundings, and the presence of the divine here and there provided a source of restraint. We cannot exploit the Earth with impunity when the things around us possess spiritual value. Whether our gods are stern or benign, whether lax or demanding, spirits in nature mean that humans are not the only things in the world that count. This process of purging nature of the divine was wrapped up in the gradual shift to monotheism that, in the Western World, occurred in earnest some two millennia ago. With monotheism, God became dis-embodied and lived apart from the physical realm. God may have created the Earth, but the Earth was not God. As our

conqueror took this first, slow step, an important impedi-
ment to full conquest was gradually removed.

A second needed step for the maturing conqueror was
mentally to separate humans from the rest of nature. When
humans are kith and kin to the animals surrounding them, it
is harder for them to kill animals for any or no reason. When
we sense some family ties, when we sense that other species
are moral actors in their own right, we cannot envision them
in purely instrumental terms. To treat nature as an object, we
must develop a sense of dualism; we must develop within
our subconscious, so deep that we are unaware of it, a sense
that humans are distinct from the rest of the Earth.

In western intellectual history, this people–nature split
emerged slowly, and the story of its emergence is a long one.
Plato had something to do with it, as did other early philosophers. Plato's theory of forms suggested that the physical
things we saw and touched were actually unreal. They were
imitations or shadows of real (that is, ideal) forms, which
existed in a metaphysical sphere somewhere. Plato cheap-
ened physical things by calling them unreal. By doing so, he
pushed along the dualism that depicted humans as special
and superior to the rest of creation.

If Plato stands for the second step in our conqueror's
intellectual growth, Isaac Newton stands for the third. It was
not enough for our conqueror to remove God from nature.
What was needed as well was to remove the inherent mystery and mystique that had always been an important part of
nature. Early humans were perplexed and dazzled by much
that they saw around them. They sensed that the Earth was
far beyond their comprehension; that it was full of surprises
and unknowns and simply inexplicable events. Early Greeks
and Romans chipped away at this sense of awe, but it took
the more concerted effort of the Enlightenment to make serious progress.

Newton was a religious man, inclined to believe that God
was at work in the world around him. Yet he set out

nonetheless to explain physical events in simple, mathemati-
cal terms. The intellectual spirit that he embodied encouraged people to view the entire Earth as a huge machine. Its
parts were many and the interactions complex, but ultimately, or so it seemed, humans could gain a full understanding
of it. Astronomer Johannes Kepler captured this idea a generation or two before Newton. "My aim," Kepler asserted, "is to
show that the celestial machine is to be likened not to a
divine organism but to clockwork." To Kepler and Newton,
the Earth was not a place of great mystery; it was only a
place of great complexity. The mysterious stimulated awe
and restraint; the complex stimulated only inquiry and
manipulation. With enough cleverness, humans could com-
prehend the ecomachine and put it to work.

One problem with early science was that it was slow in
providing answers and it left many gaps. Gaps mean ignorance, and ignorance can make us pause. In time our conqueror would deal with the nagging effects of ignorance by
developing a method of thinking that based decisions on
the known and largely ignored the unknown. This was the
conqueror's fourth step, and we can conveniently attach
René Descartes's name to it, although he hardly deserves full
credit.

Descartes was a resolute man, and the object of his dedicated quest, his grail, was a way of knowing things with true
certainty. He went about gaining knowledge through his
now well-known method of systematic doubt. Descartes
pushed aside all that he could not prove—which was more
or less everything—and began simply with the fact of his
own existence as a thinking, doubting being. With this fact as
a given, Descartes proceeded logically to prove the existence
of other things, always basing his logical chains on what
he knew.

The spirit of Descartes is very much alive among scientists when they exhibit their typical reluctance to talk about
the world's many mysteries. Their opinions are based on
what they can prove, and their tendency then is to fall silent,

even while they keep looking for more knowledge. So far as it goes, this attitude is fine. But by answering questions based entirely on what can be proven, we end up ignoring the gaps in our knowledge. When it comes time to act, we are left with no means to factor in the mysterious. Indeed, the implicit suggestion of the scientific method is that things we don't know and can't prove don't exist. When we are forced to acknowledge an unknown, one option is just to chalk up its cause to chaos—a technical-sounding term that denies the existence of a pattern, when in fact the pattern might be one that is simply beyond our comprehension. Another approach, of course, is to admit ignorance and then figure out how to cope with it.

If our conqueror had been a scientist he perhaps could have handled this limit on the scientific method. But the conqueror that has blundered his way up to the present was, by and large, no laboratory scientist engaged in a quest for pure knowledge. Our conqueror, instead, has been one who acts. Descartes, Francis Bacon, and the scientific method implicitly encouraged our actor to forget about the mysterious. They encouraged him to assume that the unknown presented no problems. Primed in this way, the conqueror could readily assume that an unknown plant had no value—not unlike his descendant today who blithely assumes that a strange chemical does not harm until proven to do so.

The era of Descartes and Newton brought a new emphasis on the importance of the individual. In the spirit of the Renaissance, man became the measure of all things. The secular Enlightenment extended this thinking, and the individual came to be the prime locus of value. Natural-rights thinking merged with economic liberalism to give the individual unheralded powers, powers that came at the expense of organic social visions. This move to embrace the individual was our conqueror's next rung on the intellectual ladder—and a big step it was.

The Puritans who stepped ashore in early Massachusetts came to develop in the New World a model community, a city upon a hill. In their model vision, they expected the individual to bend his interests to those of the community, and land-use restrictions, like social rules, were part of the vision. By the time of the Declaration of Independence a century and a half later, this community vision had greatly eroded, in New England and elsewhere. The talk had turned instead to the rights of the individual. The Declaration spoke of the inalienable right of the single human to pursue a personalized version of happiness.

Individualism in fact took centuries to gain ascendancy, and its rise coincided with the other steps that our growing conqueror took. Individualism was aided by the popular emergence of economic thought. Government in nineteenth-century America sought to unleash the economic energies of the people, and it did so in part by sanctioning individual greed. Government widened the playing field so that each player could look after himself and grab what he could. Over time this focus on the individual, on personalized rather than shared visions of the good, came to dominate political as well as economic discourse. Individual freedom became the rallying point, and both the liberal on the far left and the libertarian on the far right could join in the cry.

Once our conqueror embraced individual freedom, the conquest of the Earth could accelerate. As individualism took charge it pushed aside earlier social visions that subordinated the individual to the good of the community. With individualism in his pouch, the greedy, energetic conqueror was free to grab as he wished, without thought for the health of the whole or the community's future needs.

At this point, only one intellectual step remained for our adolescent conqueror—the step of stripping nature and the Earth of all residual special value.

In its full flower, individual liberty meant that each individual could determine what was good and bad. Value became a personal, subjective factor, something soft and pliable. In this environment, economic theorists stepped forward with a new, hard, factual version of value. What economists had to offer was the idea of market value, and they would press their idea with vigor, particularly in the arena of public policy. The value of a thing, they claimed, was reflected in its market price. Only market price was an objective standard; only market price was solid and defensible.

Influenced in this way, our newly matured conqueror began to think solely in economic terms. Cost–benefit analyses came into being, as did talk of nature as resources. Resource thinking encouraged the conqueror to break the Earth into component parts, for only discrete compounds and pieces had market value. With resourcism—seeing nature as a collection of actual or potential resources—the whole was nothing.

Resourcism brought another effect as well. As the conqueror embraced market value, intuition, emotion, and passion no longer played a role, except as they were translated into dollars. The individual consumer might still harbor a passion for birds and clean water, but that passion was personal, subjective, and, hence, irrelevant, except as it led to market transactions. Only the market had a voice.

These, then, were the intellectual steps that the maturing conqueror took on his path to the present. By the end of the journey, the conqueror had become a modern adult. As he surveyed the Earth, he saw an orderly, comprehensible scene. God was gone, as were the mysteries, legends, and special vibrations of nature. Humans were set apart from the Earth, and communal needs had come to mean nothing. Man was the measure of the Earth, and money was the measure of man. Man was the subject; the Earth was the object. With this mental framework, the conqueror could take full advantage of the new technology. With this framework, he could now tinker with the Earth in grand, unheralded ways.

The Conqueror's New Clothes

In sketching the intellectual growth of this conqueror, the aim has been to depict, not a real person, but an image of pure domination, a distillation of the idea that, to varying degrees, has guided our use of the planet in recent generations. This image shows no recognition of environmental degradation and offers no solution to it. It is, we might say, a baseline. As we'll see, the conqueror is very much alive today. But he has smoothed some of his rough edges and is dressed in far more respectable clothing.

As our conqueror rounded the end of the nineteenth century, he encountered growing resistance. There were a few critics like Albert Schweitzer who expressed heretical ideas about science and human powers. "As we acquire more knowledge," Schweitzer observed, "things do not become more comprehensible but more mysterious." This was a direct, undeniable assault on the conqueror's mastery. But Schweitzer and those like him whispered from far out of the mainstream. They were not a challenge; only the faint murmurings of one.

In the United States, the real challenge at the beginning of the new century came from the Progressive–era movement know as "conservation,' which was headed, at least in the public eye, by a lanky, aristocratic forester named Gifford Pinchot. Friend and close advisor of Teddy Roosevelt, Pinchot believed that we should conserve our forests and other natural resources. By conserve, he meant that we should manage the resources actively so that they remained productive for humans in the long run. No more raping, plundering, and then running, Pinchot announced, as the timber companies were then doing in Michigan, Wisconsin, and Minnesota. We would manage the forests so that they produced an even flow of timber in perpetuity.

Pinchot's conservation represented a challenge to the conqueror image. Pinchot thought of the future and of the health of the land. He wanted to analyze and study before acting. Yet, as challenging as his image was, as alarming as he was to the slash–and–burn timber companies, Pinchot and our conqueror shared a great deal of common ground.

Pinchot contested only small parts of the worldview that our conqueror had spent so many centuries putting in place. He saw nothing divine and little that was mysterious in the woods that he called home. For Pinchot, the forest existed to serve humans, and he evaluated it in utilitarian, instrumental terms. He was unafraid to pronounce a species worthless and to take the axe to it without delay. Man was still the subject; nature was still the object.

Believing firmly in the ability and right of humans to manipulate the Earth to serve their needs, Pinchot was confident in his planning. Like the conqueror, he placed his faith in science to unravel the complexity and help chart the most utilitarian path. Interestingly, Pinchot's faith in science was both more deliberate and less complete than was the faith of our conqueror. The conqueror plundered and ran, leaving the scientists to come in later and figure out what to do. Pinchot invited the scientists to participate from the beginning, and he became a scientist himself. He saw ahead to the problems of excessive harvesting and wanted scientists to help immediately, not trusting their ability to cure the problems later on. Once scientists formulated plans based on what they knew, Pinchot's foresters would carry them out.

Pinchot intended to be farsighted as well as scientific. He intended to consider the needs of the community, present and future; he called for wise use, and his plan was to leave a productive Earth for those who came later. Like today's utilitarian philosophers, he believed that he knew all he needed to chart the most productive path into the future. For Pinchot, as for our conqueror, human cleverness could be trusted to save the day.

The pure conqueror image does not deserve a great deal of serious attention these days, except as a way of unraveling our cultural and intellectual problems. Pinchot's conservation image, however, is of a different order, despite its considerable overlap with the ethic of domination. Pinchot, we might say, brought our conqueror into the twentieth century by tempering his zeal. He gave the conqueror new clothes and thus made him more respectable in the eyes of a public that was slowly awakening to the land's savage wounds.

Pinchot's image, like those we have considered so far, offers a prescription for our environmental woes, a prescription similar to the philosophers' magic calculator. And it is a prescription that remains very popular today.

The Voice from the West

While Pinchot was combating the exploiters from his offices in Washington, D.C., and New York, a more shrill, lonely voice was making its way east from the Sierra Nevada Mountains. Bearded, disheveled John Muir, John of the Mountains, was preaching a far different gospel of repentance. He was baptizing with a purer spirit, and he was gaining his converts.

Bringing Muir into the picture helps us see more readily how little Pinchot really intended to change our conqueror's notion of the world. Muir's challenge was far more dramatic, for Muir sought to undo much of what our conqueror had learned. When Pinchot sought to express his ideology of scientific management, he seized upon the term "conservation." What John Muir preached was the far more demanding doctrine of preservation, which meant leaving the land untouched.

Muir was raised by a stern Scottish Presbyterian father, and if he rejected much of his father's liturgy, he carried on his father's evangelical zeal. He lived frugally, on dry crackers and tea (or so he said), and his ethic was equally stern. For Muir nature was valuable on its own terms, not because it could serve humankind, and leaving a wild area untouched was very often the highest and best use. He saw strong evidence of the divine and the spiritual in the world around him, particularly in the special natural places of the Sierra Range and Alaska.

As Muir climbed the towering Douglas firs to get closer to the sky he felt at one with the natural world. He felt strong links with the Earth and was prone to deny the human–nature dualism of the classical mind. Muir's roots drew sustenance from the Romantic tradition, the tradition that rebelled against the effort to wipe nature clean of its messages and meanings. There was mystery in the hills, invisible vibrations and symbolisms in the quaking leaves. Nature was bigger than humans, Muir sensed, and far beyond the human mental grasp. It was too big for the mind; only the open heart could take it in.

Perhaps above all, Muir responded to nature in passionate terms. His talk was not of market values and minerals and board feet; it was of the spiritual and religious senses that only nature contained. Muir loved the Earth, and he wanted others to love it also. He had faith in science and respected the ingenuity of the tinker—he was, in fact, a skilled tinker himself—but he believed, passionately and vocally, that the scientists and the exploiters had done more than enough. It was time to draw the line and to place the untouched off limits.

Like Schweitzer, Muir offered only the beginnings of a challenge. When Congress created Yellowstone National Park in the early 1870s, it acted only after being told that the area had no economic value. Why not, then, preserve the beautiful? When John Muir years later pushed for the creation of Yosemite National Park in California, economic uses were very much at issue. Muir pushed, Robert Underwood Johnson and others pushed, and the park became a reality. But Yosemite would be the exception. Muir's ethic would need to await a later day before gaining broad acceptance. Even the Yosemite exception would be cut back in Muir's lifetime. Muir's last big crusade was his effort to keep the dam builders from flooding part of the new national park. He lost, and Hetch Hetchy Valley was put to use supplying water and power for the demanding cities of San Francisco Bay.

The Conqueror as Conservationist

If the conqueror image conveys a central message to us today, it is that human cleverness is now sufficient to destroy the Earth, at least as a suitable home for human life. It is a sobering, frightening power, a power that in some way needs to be restrained, a power that surely must carry equally forceful obligations. In its destructive power, human cleverness is like a loaded handgun that we bring into the house thinking that we may need it for protection. Even believing that the gun has its rightful time and place, who cannot sense the need for limits, who cannot sense that, by bringing the gun into the house, we take on obligations we did not have before?

The Earth is our house and human cleverness is the handgun. We own the gun, and now that it is in the house, it is our responsibility to control it. Cleverness is not inherently bad; it has, indeed, brought many joys and must be part of our communal salvation. But as we take up the tool of cleverness, we must take up also the accompanying obligations of restraint.

With this in mind, we can turn to consider some of the traits and difficulties of the twentieth-century version of the conqueror vision, the version that Gifford Pinchot decked out so impressively in the new conservation garb.

In his quest to promote conservation Pinchot sought to do what was right for the nation in the long run. In searching for what was right, he wanted the most accurate information at his disposal. What he didn't want, what didn't fit into his calculations, was any sense of the divine and the mysterious in nature. In this secular age, talk of the divine seems out of place to many people. Talk of the mysterious can also seem out of place, for it appears to suggest a certain disrespect for the long-term prospects of science.

Yet what talk of the divine and the mysterious helps us do is incorporate into our actions some recognition of the limits on our present knowledge. These words are ways of saying that we don't know all that there is to know, and perhaps we never will. We don't know how nature works, at least not very much of it. We don't know what will have value to us in the future and what will not, even assuming that human utility is going to be the test. We don't know, really, which actions of today will strike future generations as good or bad, wise or foolish.

Pinchot's conservation ethic made no room for possibilities of this type. Pinchot worked for the bully President, a man of action who knew what he needed to know. Pinchot, too, was decisive. His task, as he explained it in the title of his autobiography, was to break new ground, and once he chose a plow and selected a field, he went right to work. His thinking offered no way to incorporate large chunks of ignorance; it left no room for error.

Pinchot believed that he had found a way to produce high-quality timber in perpetuity, and he managed the National Forests on that assumption. In fact, his goal was more elusive than he realized. His forestry techniques were far better than the slash-and-burn techniques of the day, but he found no miracle means to eliminate the soil erosion that accompanied harvesting. He found no way to restore the soil to its original fertility. Nature's way was to allow trees to die and rot in place. Once trees were removed, the cycle of death following life was interrupted. Pinchot's methods couldn't bring about perpetual yields.

But even if Pinchot had been right in assuming that he had found the magic of perpetual yields, he made a grave mistake by focusing on timber alone. As he sought to produce timber he ignored the productive yield of vast numbers of species and the overall health of the land. Pinchot sought to turn his forests into tree farms. He didn't entertain the thought that later generations might wish the land had been managed for other yields. Ecosystems were being disrupted if not ruined, which meant that the land was losing its ability to produce many things. Pinchot didn't think that he might be sowing the wrong seeds; he didn't consider that, even with the greatest of foresight, he couldn't know exactly what to plant.

Like the conqueror, Pinchot placed his faith in science. He did so, and we do so today, in two principal ways. Pinchot believed that he knew all that was needed to make the right decisions. He knew which species to cut and which to let live. A poignant illustration of the folly of such certainty was offered by Aldo Leopold, an ardent Pinchot disciple in early life, in his moving essay collection, *A Sand County Almanac*. In a splendid essay, "Thinking Like a Mountain," Leopold described his utilitarian calculations for ridding the Southwestern National Forests of all wolves and lions so as to produce a Pinchot-style continuing yield of deer for hunters. Only later did Leopold realize that predator elimi- nation means overpopulation, which means habitat degrada- tion, which means overall decline. In true Pinchot style Leopold thought he knew enough to tamper, and he was wrong.

The second way that we can have faith in science is by knowingly leaving a problem for resolution at a later day. We build millions of gas-hungry cars and then proceed to con- sume all of the world's petroleum supplies, counting on the fact that our scientists will come up with a solution. We build reservoirs knowing that they'll silt up over time and trusting that scientists will find a way to deal with the mess. Perhaps we can best capture this second type of faith by our most dramatic modern instance of it—the huge, sealed container of highly radioactive waste, set aside "temporarily" while we search for a place to dump it. Since the 1950s, we've amassed huge amounts of this waste, knowing all along that we have no safe disposal solution. In this instance, faith in science has taken the form of a very risky bet.

If we can capture human cleverness with the image of a loaded handgun in the house, perhaps the sealed radioactive waste container best expresses our faith in science. Faith cre- ated these great boxes, one of the sorry inheritances that we'll bestow upon generations that follow.

Before leaving Gifford Pinchot, we might take up a final point about this important clash between conservationists and preservationists, the clash that Pinchot and Muir thrust upon the public stage. To a surprising and perhaps harmful extent, Pinchot and Muir have defined the twentieth-century debate over natural resources policy: conservation on one side; preservation on the other. Either we manage for high yield and use as we need, or we set aside and place off limits. For decades, Pinchot was the clear winner in this contest. In water resources policy, for example, conservation became the rage in the early part of the century, and conservation had a precise meaning. To conserve water meant to manage water flows with enough intensity, with dams here and ditches there, that not a single drop flowed unused to the sea. As this was done, water quality, soil quality, and wildlife populations all began a sharp decline, and all in the name of conservation.

The conservation–preservation dichotomy has a familiar ring after all of these decades. So strong is the familiarity that these two terms seem to define our principal alternatives. We choose one side in this instance; we choose the other in the next. In fact, though, these positions are widely separated, and in between them lie many intermediate ones. Muir's preservation ethic sought grave changes, and he staked out a spot a long way from Pinchot. Muir sought not only to get our conqueror to unlearn all that he had learned but to force him to get off the land, to leave it alone, and to cast aside much of his technology.

Muir's reaction was extreme, even though his prescription for wilderness preservation had considerable merit in many settings. But, as we saw in the last chapter, wilderness preservation is not the model for humans to use in building harmony with nature. We can learn from nature, but we somehow must apply it to a setting in which we can live.

As we develop a new way of imagining and, hence, living with nature, we'll need to steer a course between Pinchot and Muir. These towering early figures staked out the pole

positions. In the end, we're likely to head down an intermediate path.

The Promise of Ecology

The one science that seems most deserving of our faith these
days is ecology, which is the study of how species interact
with their environments. Its focus is on the totality of relationships and dependencies, and it brings the bits and pieces
of nature together like no other science does. This scientific
field is the most exciting one for environmentalists; as we've
noted, it's had considerable influence on environmental
ethics. Ecology is the one field that takes on nature as a
whole, in all of its intricacy and interconnections. An ecologist studying a raccoon is interested in everything that the
raccoon touches and in everything that touches the raccoon.

Ecology is at the heart of the modern concern over the
environment, and ecologists have been as prominent as any
other group in identifying our mistakes. It's hard to imagine
that our consciousness over environmental problems could
have come as far as it has without this important scientific
branch. Aside from the masses of scientific data that ecological study has produced, it has yielded several splendid
visions.

Ecology has shown us some of the complex ways that
species interact and depend upon one another. Early scientists largely sought to accumulate information on each
species separately, and their work mostly involved chopping
up dead plants and animals. Aldo Leopold, probably the
most revered figure in American environmentalism, was an
early ecologist, and he helped pioneer the new field of
wildlife studies. Leopold examined wildlife in the wild, not
just in the laboratory. He was interested not just in taxono-
my and physiology but in behavior and habitat needs and in
the interactions of all life. Ecologists took to the field,
because they had to. Only in the wilds could they identify

behavior, food, habitat, and dangers. Only in the wilds could they study animals as living creatures, not mere carcasses.

In its early days ecology helped articulate the image of the energy pyramid, which started with broadly based layers of plants and microorganisms and progressed higher and higher as animals ate plants and larger animals ate smaller ones. At the top of the pyramid stood the largest carnivores and omnivores, including humankind. To those who thought about it, this pyramid helped show dependencies and displayed how energy flowed through the entire system.

Ecology also showed how species competed with one another and how, given time, certain species would dominate others. In the north woods around Lake Superior, ecologists explained how the jack pines, aspens, and birches were the first to colonize an open area. Over time, these trees were pushed out by the black spruce and balsam fir. In turn, these species were pushed out by the white pine, which was viewed as the dominant tree of the so-called climax forest.

As ecology has come of age its depictions of life have become less simple and, presumably, more accurate. Ecologists now realize that the whole concept of community climax is misleading, for climaxes are always tentative and subject to being upset by a wide variety of natural forces, including fire, disease, and weather. In the north woods, white pine is vulnerable to fire and disease, and when it dies, it opens the way for the jack pine and birch to return. This disease- or fire-driven transition from white pine to jack pine and birch is as natural as the shift onward to balsam and spruce. Animal populations, too, experience wide fluctuations, and in the long run, climatic changes bring even greater alterations. The idea of the climax having lost much of its explanatory force, ecologists today speak about nature in terms that are far more fluid. Some species are expanding their range at any time, while others are contracting. New forms are emerging, particularly at the microscopic level. Images of conflict have yielded to more mixed images that include large doses of symbiosis and cooperation. Increasingly, species are explained and given positions based on their relationships rather than their physical attributes. The individual organism is not so much an autonomous creature as a locus or node through which energy flows and processes take place.

It is hard to capture ecology and its messages with a single depiction. Perhaps the best we can do is select the largest, most intricate spider web that we can find. Transform the web from two dimensions into three, and then spin extra strands so that every intersection is connected with every other one. What we produce is a delicate, intricate maze of interdependence, which is perhaps the core of the ecologists' tale. Next, let us add the cast of animals: the spider, which constantly spins new threads and moves and eliminates old ones, and the occasional prey, which squirms and flaps in an attempt to escape and succeeds only in snarling many strands. Finally, we add the winds and storms.

Whenever we act in nature, we pull one of the strands of this most intricate of webs. We may know the first effect or two of our pulling—with luck, they'll well be what we want. But the initial disruptions are inevitably followed by others. We spray to eliminate mosquitoes to protect ourselves from bites. The first effect is obvious, and what we want. But what of the birds that feed on the mosquitoes? What of the seeds (particularly, from our perspective, the weed seeds) that these birds might also have eaten? What of the predators who would feed on the birds? What of the pesticide that lingers in the environment to disrupt other species or that builds up in the tissue of various animals and begins to cause unknown, untraceable mortality? The ripples expand ever outward. Ecology helps to see these effects but tracing them is hard; ecologists can't be everywhere and know everything, and they aren't and don't.

Ecology broadens and reiterates the messages that John Muir absorbed through direct observation. Pull one thread, and you see it attached to all else. Philosopher–theologian Alfred North Whitehead drew upon the same understanding of relationships to develop his influential process theology. Value for Whitehead came from interactions and dependencies, from processes undertaken over time rather than from autonomous existence.

With all things connected to all other things, all value must be shared, and all parts must be important. It was with this idea in mind that Aldo Leopold began decrying the loss of species. Man was playing the mechanic and rearranging as he saw fit. "To keep every cog and wheel," Leopold admonished, "is the first precaution of intelligent tinkering."

Ecology enriches our understanding of nature and adds vividly to the benefits that we get from our study of the wilderness image. By allowing us to see where we have not seen, it helps us uncover the barely visible scars of many environmental wounds. Exposed to the details, we see how complex everything is and how relatively little we know. Nature is one of our best teachers, and if we are to survive it always must be.

Ecology is not without its risks, and it suffers from the same incompleteness as scientific study in general. Studying the details of relationships, it is easy for ecology to lose all sense of the whole. By being precise in our factual observations, we can cast aside the kind of emotion and wonder that probably attracted us to the study to begin with. Much of our sense of intrinsic value in nature comes to us in intuitive, emotive ways, and if we're not careful, ecology can cut us loose from our basic understandings of why things are important. We become focused on the detail, on the quantitative, and lose sight of the context and value.

Ecology has filled some of the gaps in our knowledge, but it has probably exposed more ignorance than it has removed. It hasn't eliminated the need for us to appreciate the strangeness and mystery of the natural world. It hasn't altered the need for us to reorder our minds and discard much that carries over to us from our days of pure conquest. Ecology will make the task easier, but the pruning, the mental reshaping and spiritual growth, is something that we each still must do.

7

Embracing Our Ignorance

God guard me from those thoughts men think
In the mind alone ...

—William Butler Yeats

In many cities and towns dotting our nation's landscape, the ecological issue that grabs the most attention these days is the issue of garbage. Americans generate garbage very well, better than the residents of any other country, and it shows. Our landfills are filling and closing, and few new ones open. Semitrailers haul eastern garbage to the Midwest. Trucks everywhere carry garbage from cities into the countryside. Railroad cars and barges are getting into the act. What are we to do with this waste?

My home community is a good example of this festering problem. It has struggled for years with its garbage and, given the glacial rate of progress, the problem will probably be around a long time. The issue isn't staying alive because of inattention. To the contrary, the attention lavished on it, by government officials, local environmentalists, affected neighbors, students, garbage haulers, farmers, editorial writers, and the like—everybody, it seems—is unprecedented.

After a brief interlude of calm, some new wrinkle develops—some decision is about to be made, or some new report or plan is or isn't released, or some vote is to be taken—and then the placards go up, the pens are dipped in acid, the flyers are printed, the emotional appeals are drafted, the meetinghouses are packed, and on and on the commotion goes.

There are times when I wish Shakespeare were here. He'd know how to write about the foibles and intrigue of popular government and put it all in perspective. He'd pick out his Lady Macbeth and John Falstaff, his Puck and Mistress Ford. He'd be just the one to have some modern-day John of Gaunt wax eloquent about a sun-dappled vale soon to succumb to a forty-acre landfill.

This local garbage problem doesn't drag on because some mischievous cabal takes pleasure in enraging the populace. It drags on because the issues are difficult and the possible solutions are uncertain, unpredictable, and expensive. Uncertainty, not perversity, is the taproot.

Local government has tried repeatedly to site a new landfill only to find in the end some unexpected environmental condition that proves disqualifying. In analyzing possible sites, local leaders don't know what things people will throw in the landfill, whether and when the landfill will leak, how quickly the waste will decompose and into what, where the leachate will flow and how fast, what the migration patterns will be, what groundwater will be affected, what surface waters might be harmed, what wildlife will be disrupted by the pollution, and so on. Incinerators have risen and fallen in popularity, not for the first or last time I suspect. With this option, we don't know what the burnable waste flow will be, how successful the screening will be, what the ash will be like, what the exhaust will be like, where we can put the ash and the screened material, what the costs will be, and so on. Composting has also appeared on the horizon, only to raise similar questions: will we screen the waste, will it degrade on schedule, what will the resulting compost be like, can it be put on fields without endangering humans or natural health.

At one thousand square miles, Champaign County has many sites for whatever new facility is chosen. Of the thousands, which site is the best?

Farm groups don't want prime farmland lost to a landfill, and nearly the entire county is prime. Look outside the county, they urge. Environmental groups and others push for recycling collection operations, bottle bills, mandatory recycling rules, volume–based garbage collection, waste–reduction measures, and the like. These, too, are loaded with uncertainties, about economics, popular reactions, and trends in the waste flow, to name just a few.

My sense is that ignorance is the driving force behind much of this activity. Fair–minded officials have waded into a problem knowing perhaps one-tenth or one-hundredth of what they would like to know come decision time. The known facts are many, and a goodly number of participants, including some of the most vocal ones, know few of them. Even all of the known facts are distressingly slight in comparison with what we ought to know, which in turn is far less than what there is to know. The whole topic is buried under layer after layer of unknowns, to which we then add the normal factor of human frailty.

The ignorance here is not the ignorance of the lazy or the slow. It is the ignorance of the decent, motivated intellect that is honestly and legitimately overwhelmed. Yet, in the end, decisions need to and will be made. A landfill will be added, over some poorly known geologic formation, to accept a dimly perceived waste flow, to be followed by a poorly understood process of decay and perhaps leakage and perhaps areawide environmental contamination. If an incinerator is built, if composting is done, if recycling processing is undertaken—all will be done in the same way.

If the garbage issue in my town offers a particularly vivid illustration of problems that can perplex and befuddle, it differs from other issues only in degree. Ignorance is a regular visitor in all of our lives, and in all seasons, at all places, it arrives to play an annoying, unpredictable role.

Assessing Ignorance

Throughout this inquiry, we have confronted the limits on human understanding. Before proceeding, we need to respond to the up–front objection that many people will raise: How much ignorance really remains today when our libraries and data banks are bulging? Don't humans know far more today than they have ever known? With hundreds of scientific fields and subfields, with experts in each, hasn't our knowledge simply exploded? If primitive tribes could live in equilibrium with nature with their far less impressive knowledge, can we not do the same?

We can begin piecing together a response to these questions by considering the types and forms of ignorance that appear in daily life.

1. Some things humans simply do not know. This is the most obvious form of ignorance, and the one people have in mind when they point proudly to scientists and libraries. Two types of ignorance make up this category. In some cases, we are aware of a knowledge gap and, hence, aware of the ignorance. In other cases, we do not see the gap and are ignorant even of the ignorance.

2. Some things we could learn with our scientific tools, but we haven't taken the time and spent the money to do so. With enough effort we could test every groundwater well for traces of thousands of chemical compounds. We could test them once a year, once a month, once a day. We could investigate the leaching patterns and potentials of every landfill and every underground injection well. We could, but we haven't, and ignorance retains the upper hand.

3. Some things are known by someone somewhere, or the knowledge is in a book or computer, but it isn't in the active consciousnesses of the people who need it when they need it. Here we are talking about active, useful knowledge. What a particular manufacturer knows about the ill effects of a pesticide is of little help when the information is stamped Trade Secrets and locked in some filing cabinet. What the

ecologist knows about the cornfield is not what the farmer knows.

These three categories of ignorance are set forth in what is probably their ascending order of importance in our daily lives. The first category is likely the least important, despite its considerable vastness and the prominence that we attach to it. And within that category, known gaps might well be less important then the unknown ones.

A bit of reflection on these categories should convince us that our ignorance is vast. But we still might wonder: Is it shrinking? As we learn more are we expanding the side marked Known and shrinking the side marked Unknown? In time, will our knowledge reach the point where ignorance is a modest problem?

These questions, perhaps surprisingly, are not easy ones to answer. It is undeniably true that as a species we continue to learn new things. Much of the new knowledge is highly specialized and, to the extent it reduces ignorance, it does so largely in the first category. Knowledge, yes, is expanding.

However, to a degree typically unappreciated, human knowledge—even of the first type—is also rapidly contracting. Even now, most knowledge is not written down. It arises from experience and is passed on orally, if at all. Much of the most important knowledge is local knowledge, particularly in the case of knowledge about the land and nature. With the human exodus from the farms, our countrysides have lost, and continue to lose, enormous quantities of knowledge. Kentucky farmer Wendell Berry makes this observation in many of his writings, and he does so in focused, particular terms. Berry was born four miles from where he now farms. Over the decades, he has gained a special intimacy with his small farm and the neighborhood. He is a repository of unwritten local knowledge.

At one time, for example, people in each locale knew how to reduce crop losses to weeds and insects without resorting to expensive, dangerous, energy-consumptive chemicals. The tricks differed widely with the variations in the land and the ways it was used. This knowledge arose slowly, through trial and error, and was but a small part of the local cultural tradition. Many farmers knew their small farms with Berry's type of intimacy—what each acre had done, what it could do; how it could and couldn't be treated in the long run. The tricks and crafts and simple tools were all part of this knowledge. In the farmhouse, in woodlots and gardens, in towns and cities, there were and are equivalents. Much of this knowledge in rural areas is gone or going, and its loss should be a cause of alarm.

We can abandon much of our departing knowledge with no ill effects on our dealings with the land. But then there is other knowledge—the kind of knowledge that, when put into action, distinguishes a land steward from an owner who is indifferent to the peculiarities of place. There is the kind of knowledge that is swept away when we think of land as an abstract legal parcel, or as a permanent economic asset, or as a tool with a specific dollar output. There is the kind of knowledge that is too often unknown and irrelevant in the offices of distant corporate leaders who report in purely numerical terms to scattered, distracted shareholders.

Much scientific effort today is invested in the creation of new chemical compounds. New herbicides, insecticides, fungicides, and fertilizers are added directly to the land. Other new things—lubricants, cleaning solutions, adhesives, preservatives, dyes, food additives, plastics, and thousands of other creations—arrive in the air and soil less directly. Each new creation adds to and underscores our ignorance. Each new creation raises new questions of health and safety for humans and other life forms. Wildlife species react in widely divergent ways to a single new compound, and so do we. Small doses of penicillin can kill some rodents, yet penicillin was one of the early miracle drugs for humans. Thousands of new compounds; millions of species; uncountable arrays of settings and interactions: with each change, the questions and the ignorance expand.

Another form of ignorance comes from relying on samples and averages. A food distributer announces that its peaches have only a trace residue of pesticides. The distributer means, of course, on average, and knows nothing specifically about any particular peach. The distributor announces that trace residues are harmless. He means, however, in the short run and on average, not for any given person.

Around 1948, at the beginning of the age of synthetic pesticides, the United States used some fifty million pounds of insecticides and lost seven percent of its preharvest crops to insect pests. By the mid-1980s, pesticide use had risen to six hundred million pounds. Thirteen percent of preharvest crops were then being lost—nearly twice as much. Many developments lie behind these statistics, in terms of land use, crop patterns, tillage practices, and insect mutations. The essential fact is that, with decades of research, the farm pest problem is not getting smaller and our ignorance is not receding. The research task, to be sure, is a hard one, particularly when insect species develop tolerances and powerful insecticides are restricted because of unacceptable side effects. But nature is like that and is not soon going to change.

The old saying is that we don't know what we don't know until we know it. It takes an act of faith to believe that, on balance, our ignorance of the natural world is receding rather than increasing, to assume that ignorance will not continually slap us in the face with the effects of our miscalculations.

If ignorance will always be with us, how can we deal with it?

The conceptions and images that we've looked at address ignorance in varying ways. The economic image of externalities denies that ignorance exists. It assumes that we can trace, calculate, and internalize everything. The magic calculator image sinks its roots into the same artificial soil. The accounting image of land's permanence and the law's principal images avoid the issue in a different way. They ignore the little evidences of decline, deterioration, and interdependency; they assume that even well–known facts do not exist or that they are communally irrelevant.

Containing Our Ignorance

One way to respond to ignorance is to fight it by seeking more knowledge. This has been and will need to be one of the central responses, and it is not discredited simply because we admit in advance that ignorance may outpace our learning. Studies inevitably should focus on the latest evidence of our mistakes—on the dead fish, the declining bird counts, the stunted trees, and the other facts that seem to greet us at each turn. If we are wise, we'll also fund work that is designed simply to learn, perhaps to learn about land health, perhaps to turn up undetected evidence of illness— the choice of unknowns is endless. Some of the work should consider issues of practical importance to everyday living, issues of households, yards, and businesses, how our little acts of daily living can operate more gently.

If the second and third categories of ignorance are more important than the first, then most of the effort combating ignorance should be aimed accordingly. Let us study important aquifers before pollutants turn up, not after they've done so. If we are going to decrease ignorance in everyday life, far more information will be needed on labels and on public disclosure forms. Each product should contain instructions on safe disposal as well as on how it can be used. In each community we need to inject practical, preventive knowledge into the daily lives of community members. File cabinets and journal reports contain hard–earned wisdom on solar energy and energy-conserving construction techniques. There the wisdom largely sits, little influencing most home designs. Economists tell us that the market in time will react as energy prices rise. But are we condemned to be purely reactive when a new house today should last a century or more? Today's housing codes do not let the market dictate responses to human health risks. Is the health of the Earth less important?

An indispensable component of this educative process will lie at the individual level. Humans by and large dislike imposing harm, and their ethical impulses typically arise in response to perceptions of harm. The more acute our perceptions of nature, the more likely we are to spot environmental degradation and to sense that something is morally wrong. As the prairie, the wetland, and the northern lake each gain a more particularized focus, we are more apt to spot damage when it occurs.

What is needed is self-teaching, and of a local kind. We must become more sensitive to the nature that surrounds us and take an interest in its health. Some knowledge will be book knowledge about species and ecosystems and interactions. Some knowledge will be more local. It will be the kind of knowledge that typifies sensitive place people—people who are as concerned about where they are as about who they are with and what they do. As we gain more knowledge, we must be quick to challenge claims that indispensable knowledge can somehow be privately controlled for economic gain.

This possibility reached the United States Supreme Court in a 1984 controversy, and the Court, happily, handed down a ruling sensitive to the community and the Earth. Monsanto Co., a pesticide manufacturer, claimed that it should have a constitutionally protected property interest in information about its new pesticide, including health and safety data. The federal statute on pesticide registration required the manufacturer to disclose this information to the Environmental Protection Agency (the federal regulatory agency) and gave the EPA the authority to turn it over to the public. In the Court's opinion, no private property was appropriated by the mandatory disclosure rule, at least in the case of information given to the EPA after the federal statute took effect. The public's interest was the greater one.

It is, surely, mean–spirited and irresponsible for a company to assert this kind of private interest, for a company to claim that it can develop a compound designed to kill some form of life, to push the product on to the market with enticing, selective disclosures, and yet withhold much that people need to know about its environmental effects. As a community, our movement must be in the other direction. Those who make and sell must tell all they know and do so on labels or in other highly usable forms. Manufacturers and sellers must accept responsibility for a product after it leaves their hands—responsibility for the health of the users and for the health of the Earth. In designing and manufacturing goods, vendors should investigate issues of safe use and disposal and should tell the users about what they have learned.

As we combat ignorance by gaining and disseminating more knowledge, we must be prepared to act based on probabilities and possibilities, without demanding certainty. On issues like acid rain, global warming, and ozone depletion, certainty is hard to come by. It is foolish to push on indefinitely in the belief that problems exist only when we can prove their existence and prove their cause. We have simply seen too much evidence to the contrary. We need not assume the worst, but we must make assumptions, and we should use intelligence, foresight, and sensitivity in deciding what to assume.

Assumptions will also be needed when we contemplate the larger context in which we act. When we go camping in the wilderness, we leave the campsite in the morning not knowing who will use it next. As a matter of strict logic and certainty, we can't say whether the next users will want the campsite left neat and clean, whether they'll want firewood gathered and stacked, or whether they'll want the area restored to a natural look. The next users are not around to tell us, and we can only guess at their preferences. This kind of uncertainty is what we face when considering the issue of duties to future generations and to other species and the Earth.

In thinking about this hypothetical wilderness campsite, we could take the route of some observers and claim that, lacking certainty, we cannot develop a clear sense of duty. Because we can't be certain of the details, we may assume that no duty exists. But in fact this route doesn't eliminate the need to make assumptions; it simply changes them. When we deny the duty, we assume that those who come later have no preferences that can be upset, which is surely an unwise assumption.

Fortunately, many real–life wilderness campers don't need this kind of artificial certainty. There is a wilderness ethic based on assumptions, and most campers abide by it, whether or not a purely rational camper would do so. We don't know but strongly sense that those who camp next will want a clean site, undamaged by our visit. We sense that cleanliness and undisturbed nature are things that count and that they will continue to count. Generations that come after us will most likely want clean air, clean water, and fertile soil; they will want to see and study virgin forests, flowing prairies, and rich riparian ecosystems. These are not difficult assumptions to make.

The same issue of faith and certainty arises in dealing with other species. We don't know the value of most other species, either to humans or to the ecosystems of which they are a part. Our assumption until recently—not a wise one—has been that a species is worthless unless and until it has proven its worth by fetching a market price.

We can study and learn, we can assume and guess, and we can still make grave mistakes. If we have an ability to learn, we may avoid Gifford Pinchot's undue emphasis on timber production and Aldo Leopold's underappreciation (in early life) of the lion and the wolf. But if the past is any guide, we'll commit our own contemporary equivalents, and many of today's tools of destruction are more powerful than Pinchot's saw and Leopold's gun.

To go further, to develop a sustainable, healthy link with the Earth, we'll need to alter the frame of reference and make changes to the decision-making processes. If errors are inevitable, let's plan on their occurrence and make room for second chances. Let's avoid placing bets with nature that we cannot afford to lose. In evaluating risks, we cannot simply multiply the gravity of the potential harm by the probability of its occurrence. At some point, the gravity of the harm will be such that we cannot afford to lose and therefore shouldn't take the chance. A healthy respect for human ignorance is one factor pushing us to act conservatively. A similar respect for the human ability to act stupidly, and for human errors to cascade, is a second factor. In the end, some communal risks are simply not worth taking.

Nuclear power offers the most conspicuous example of the desire, at least of some people, to make room for second chances. Maybe nuclear reactors won't leak. Maybe radioactive wastes will stay contained. But these bets might just be ones we can't afford to lose. Past a certain scale, as C. S. Lewis observed, the technological choice we make is not just for us but for others; at some point, we choose for all others. Needless to say, nuclear bets seem even more risky when with a less expensive investment in energy conservation we could have eliminated the need for nuclear power plants in the first place.

In many lesser ways, we can and should prepare for second chances. Building a landfill on top of a vital drinking water aquifer can hardly help being irresponsible. Even if we could peer into the ground and steer clear of the thin cracks through which contaminants can slowly leach down, have we never heard of the earthquakes and tremors that redesign the ever-shifting Earth?

In preparing for errors and second chances, we'll likely find it helpful to develop prophylactic rules of conduct, rules set conservatively as a precaution. Prophylactic rules show respect for the Earth and its inscrutable integrity. By reducing risks, we also reduce the need for more information—which is, of course, a way of slowing the advance of ignorance.

In adding up these various ways of responding to ignorance, we begin to develop a sense that we can no longer act, as our conqueror learned to do, by considering only what we know. Ignorance, the prospect of error and failure, the need for second chances—all of these must be factored into every decision we make in our daily and communal lives.

————

My local government officials are struggling to deal with their ignorance about garbage. Some call regularly for more studies and more facts. Others, the gamblers, are willing simply to pick a number in the garbage lottery and hope for the best. Many of the participants are partisans who care far less about their yawning ignorance. What they want is to block one of the options—often the landfill in their backyards— and they don't seem to worry much about detailed comparative inquiries: the other options all look better.

Politics involves dealing with today's problems today and letting tomorrow take care of itself. Local politics especially is an infertile breeding ground for long–term social visions. The garbage problem, here and elsewhere, is not an isolated issue. We produce waste in part because we don't think much about where the waste is going and what its effects will be. This is not a garbage problem, it is a people problem, and it is not a question of ill–will or clashing fundamental needs, it is a question of ordinary inertia and insensitivity.

The long–term answer is waste reduction and near–total recycling. These changes, however, will come about only when people want them to happen and undertake to make them happen. People are the problem and must be part of the solution. But getting people involved, on this task or any other, is a long, slow, messy proposition. People buy products and use them; part of that use must one day mean proper disposal, in a way that doesn't harm the Earth's health. And proper disposal is far easier when we reduce the waste in the first instance and alter the composition of the waste flow to facilitate recycling and proper disposal.

My hope is that one day soon my community will conjure up some image of people dealing rightly with their garbage, dealing with it in ways that reduce the consumption of resources and largely eliminate the pollution of the land, air, and water. If we can create this image in the popular mind, we can then call it forth as a guiding vision. We can keep it in our sights at all times and use it as a way to respond to our ignorance. We can study the image, identify its component parts, and devise ways to start making it a reality. We can begin paring away the options and ideas that conflict with the realization of the image—the options that assume recycling is a fad, or that waste per capita can and will double, or that we can deal with the problem technologically, without getting people involved. Quick slices like these, and the options we need to consider, and the ignorance associated with those options, are rapidly reduced.

Even with the clearest of visions, local leaders will be left with issues of tactics, but what was once overwhelming may become manageable. We want a solution that does not waste and does not pollute or that does these things as modestly as possible. We can't prove that this image is the best—we don't have the facts to do this, and never will. To embrace this image, we must open our hearts. When the facts run dry, it is time for intuitions and sentiments to lead us forward.

Sentiment and Intuition

One thing that our conqueror learned on his march to modernity was to think scientifically and to act on known facts. But as my local councilmembers have found, the supply of facts can run out, and something must fill the gaps. The gaps will be less imposing when the decision-making processes contain, in addition to facts and logic, a large measure of humility, respect, and magnanimity; when they allow room, that is, for ethical visions and ideals that arise in part through sentiment and intuition.

Most philosophers in modern America develop their ethical norms in strictly logical fashion, following (broadly

speaking) the utilitarian route, the rights–based, deontologi-
cal route, or some blend of the two. Because their ethical
norms are based on logic and include as few assumed principles as possible, philosophers can argue about them at great
length. There are, however, other ethical approaches. They
are less rooted in logic and human cleverness, but they may
well be more important, both today and in the past, and
they find rising favor among many environmental thinkers.

Few humans govern their daily lives by ethical schemes
based on lengthy logical chains. For most of us, right and
wrong, good and bad, are based on impulses and intuitions
that are tempered, but not squashed, by reason. If purely
rational thinking enters the picture, it is often (although not
always) to justify something we've done or seek to do.
Somehow we sense that something is or is not right. Some
actions make us feel good; others engender feelings of guilt
or annoyance or shame. Our intuitions and feelings, to be
sure, depend in large measure on the surrounding culture.
But many people who are far out in front in advocating
expanded concepts of justice are driven by emotional urges,
often highly personal, idiosyncratic ones, urges that simply
will not be denied.

Sentiment– or intuition–based ethical schemes, in fact, are
of vital importance in daily life, most especially in engender-
ing a sense of wrongness. Logical schemes offer an important
supplement and can show how and why things that seem
right may in fact be wrong. But for basic guidance, intuition
is often the first guide. Why is kicking a dog wrong, we
might wonder? There are, to be sure, logical answers, but
they are surprisingly difficult to develop, especially when we
"own" the dog within the meaning of our narrow–gauged
legal culture. It is wrong, we answer, just because it is,
because a dog is relatively helpless and because kicking it is
cruel. It is wrong because it feels wrong, because it conflicts
with our sense of virtuous conduct.

Emotion supplies a sense that something is wrong and
the energy to do something about it. It prompts us to inves-

tigate and encourages a search for the consistent set of ethical principles that are at stake. In all likelihood, the principles are there. We can spot them once intuitions and
emotions stimulate us to value things so that the conduct we
dislike comes out as wrong.

We must be careful, of course, in imposing judgments on
other people and allowing highly charged emotions to run
wild. In discourse with others, in fact, logic may need to
assume a greater role. And when the heart is led by ego and
desire, the mind may need to apply the brakes.

But we should feel no hesitance in sharing with others
our less rational, more powerful sensations, particularly
when we sense something wrong. And when we set rules
for our own conduct as individuals, an emotion–based,
emotion–charged ethical ban is more than adequately
grounded. There will be those, to be sure, who lack a sentiment of care or who are reluctant to respond to their feelings. To these people we must open our hearts; we must
share our sensations; we must supply them with courage.
We must offer, not our short–term impulsive feelings—
which can be as dangerous as short–term thinking—but our
more reasoned intuitions, ones that we have subjected to
testing and reflection.

David Hume, the eighteenth–century Scottish thinker, is
often listed as an early pioneer of this mode of ethical
thought based on sentiments rather than strict logic.
Sentiment and intuition, however, undoubtedly provided the
dominant orientation long before Hume's day. In fact, not
until the eighteenth–century Enlightenment had pure reason
gained sufficient strength for intuition– and sentiment–based
thinking even to need justification. Outside at most a narrow
slice of the highly educated, sentiment has long played a
central role.

Sentiment's ethical companion over the centuries has
been religion, and religion to this day provides a solid base
for many ethical schemes that encompass the Earth. Our
dominant culture adopted the conqueror's mentality that

purged spirits from the natural world. But off to the side have always been godly critics. And they have become more vocal in recent years. When we sense God's presence in nature, it is hard not to feel values and limits. If nature is the handiwork of a distant God, it all was made, presumably, for a reason and according to a plan. If we cannot discern the pattern or see the inherent value, it is there nonetheless, and our duty is to respect it.

Philosophers by and large push religious arguments off to the side—not because the arguments are wrong or can be disproved but because they can so easily end the debate. Sentiment-based theories have the same characteristic. When we argue that a thing is wrong because we feel it is wrong or because God (or some other higher authority) says it is wrong, there is no next logical step. Statements of this type engender responses that are more parallel than sequential. We say in response that our own feelings are different or that we sense different divine guidance. The conversation becomes highly personal, with few sequential steps.

Yet, if philosophers believe that their value-clarifying efforts are best offered in the logical realm, those who seek only guidance for themselves need not be so limited. An ethical impulse may be highly religious. We may feel that all things in nature are inherently valuable. The redwood grove may be a divine temple, as deserving of protection as Mont-Saint-Michel or Hagia Sophia. News of oil spills can engender legitimate emotional outrage, an outrage that can quickly build in response to highly anthropocentric news stories that provide only a few selected bits of the story. The news headlines speak of humans hurt and dollars lost. But when millions of gallons of oil are spilled and two people die, perhaps the Earth's loss should deserve first mention. It is not out of the question, in fact, to think of an official ceremony of apology to the land.

The essential point is that sentiment, intuition, and religion can all offer strong bases for an environmental ethic, one that supplements what our logic tells us is wrong. All can help us identify wrongful conduct. By doing so, they help us embrace our ignorance and deal with the gaps. That which we sense is wrong we should avoid, whether or not the known facts fully support that conclusion. And as we avoid it, we often can sidestep whole hosts of problems. When ethical norms encourage us to live gently and simply, to respect nature in all its forms, to limit our needs and our pride, we are likely to find that the resulting gaps in knowledge suddenly become more modest.

Charles Darwin is known most popularly for his observations on the physical evolution of species, but he extended his evolutionary theories to ethical norms as well. Over time, he claimed, the ethical circles of individual humans have expanded outward, from family to tribe to nation and onward to all humanity. The most prominent summary of this ethical evolution was set forth by Aldo Leopold in his much-lauded essay, "The Land Ethic." Leopold told the story of wayfarer Odysseus, who hanged twelve slave girls for their transgressions in his absence. "The hanging involved no question of propriety," Leopold reported. "The girls were property. The disposal of property was then, as now, a matter of expediency, not of right and wrong." In time, conduct once governed by norms of ownership began to be viewed in human, ethical terms. As slaves were once property and are now fully human, so other elements of the natural order shall one day be embraced by our expanding ethical reach.

Through his land ethic, Leopold expanded his own ethical reach "to include soils, waters, plants, and animals, or collectively: the land." "We abuse land," he observed, "because we see land as a commodity belonging to us. When we see land as a community to which we belong, we may begin to use it with love and respect." Leopold's sense of value, which he derived from his intuitions and feelings as much as from his immense ecological knowledge, placed moral worth principally at the community or ecosystem level. Leopold was able to distill his ethical theory into a simple, powerful injunction: "A thing is right," he asserted, "when it tends to preserve the integrity, stability, and beauty of the biotic community. It is wrong when it tends otherwise." This was the heart of his land ethic, the litmus test for gauging human action. This ethic, as Leopold explained, "changes the role of *Homo sapiens* from conqueror of the land–community to plain member and citizen of it. It implies respect for his fellow-members, and also respect for the community as such."

Like Darwin, Aldo Leopold was an acclaimed, perceptive scientist, the kind of person in whom we might expect pure reason to reign. But Leopold was fully prepared to use sentiment and intuition, not instead of knowledge, but to supplement it and to fill the gaps. For Leopold, the acquisition of knowledge led to the recognition of greater ignorance, and he was not content to focus simply on what he knew. His knowledge of wildlife was perhaps greater than that of any of his contemporaries. Yet he mixed his vast knowledge with an equally vast humility in the face of nature's continuing mysteries.

Like Leopold, we can mix a growing knowledge of the natural world with an evolving, expanding sense of earthly justice. By acknowledging the legitimacy of our intuition, we can accept the testimony of our senses. We can question and perhaps condemn that which strikes us as ugly, destructive, or unneeded, particularly when it happens in the place we call home. Through intuition, we can attribute moral worth to future generations. We can sense value in intangible aggregates like species and ecosystems. In the end, it may be that we can expand the perimeter of our sense of justice only by an act of faith, by an act motivated by emotion and guided by a nagging, pressing sense that things need to be different. This is the way values arise, and this is the way that the Earth will gain protection.

In thinking about sentiment as an essential supplement to our reason, it is important to keep it as distinct as possible from the rushes of passion or desire that ebb and flow within us. *Sentiment* refers to the long-simmering senses and intuitions that become deep-seated within us, through observations and involvement in the world, and that provide the foundation for a view of our place on Earth. Our senses of beauty and of harmony, or quality in general, are based on long-developed, time-tested intuitions that are hard to set into words and impossible to prove in any sort of scientific manner. But then, who can prove that Mozart penned high-quality symphonies, that Shakespeare produced high-quality plays, or that Mother Teresa, in her charity and selflessness, is an embodiment of high virtue?

On matters such as these, logical proof is not what we are after. But neither are we after mere whim or impulse; nor are we agreeing that any individual's perspective is as good as any other's. It makes sense to talk about standards of quality—be it in music or literature or in the many attributes that make a person worthy of praise and emulation. We recognize quality only with a developed, well-tuned sense of what is good and beautiful and just. The brain plays an important role in assessing, sorting, and distinguishing. But at some point the brain's contributions come to an end, and an act of faith is needed.

To call something good is to offer an assessment that transcends individual facts and logical proofs. When the facts and proofs are left behind, the path becomes more uncertain, more debatable. The guideposts are not logical links; they are principles and ideas that have shown their worth in familiar settings and that deserve to be followed as the surroundings become increasingly unknown. Within the family—perhaps the most familiar of all territories—we know from experience and received wisdom that the best attributes are kindness, gentleness, generosity, honesty, and love. People who embrace these values we call mature adults. A sentiment-based environmental ethic takes these time-tested values from a well-known setting and applies them more broadly to a setting far stranger and far grander.

Science Well-Tempered

One of the conqueror's mental assumptions that we must prune back is the tendency to equate the possible with the desirable; the feasible with the suitable.

In the days before Gifford Pinchot, the conqueror clear-cut the north woods around the Great Lakes. Clear-cutting was possible and profitable, therefore it was desirable. During the mining-rush days, miners who had exhausted surface placer deposits soon turned to hard-to-extract mineral lodes. The desired minerals were encased in rock deemed valueless, and miners developed hydraulic extraction techniques that pulverized the rock with jets of water. Once the trace metals were removed, the rock and water rushed downhill and downstream. Streams became undrinkable, fish died, streambeds were scoured, and rock flows wrought great destruction. But it was all scientifically possible and made money in the short run; it was all, therefore, desirable.

It was desirable, that is, until values intervened to call a halt. The timber harvesting and mining were technically feasible but ill suited to the land under any sensible scheme of values. When short-term economic gain drives our conduct, the desirable collapses into the feasible and the sole question is one of profitability. When we take a longer, more inclusive view, the question of what is possible is only one of many. Other issues come forward: Is it consistent with the continued health of the land? Does it promote the health and cultural harmony of the people? Can it be sustained in the long run without damaging what will be passed on to later generations? Is it consistent with the continued functioning of a healthy ecosystem?

In dealing with nature, science and knowledge will play vital roles. But we cannot simply follow their lead. They must be viewed instead as tools, as sources of options from which we can select in constructing a good, moral life. We must recognize that those who possess the science and the knowledge will often be guided solely by short-run economic concerns, which means that, like the timber companies and miners, they will often equate the possible and the desirable. And the economic siren will always be alluring, with talk of jobs and competitiveness amid the natural destruction. One indispensable role for science will lie in alerting us to dangers and damages—making us aware, as Leopold expressed it, of the Earth's nearly invisible wounds. When wounds warn us of deterioration, change will be needed, and science can help find alternatives that are more gentle. Science can help identify risks and quantify them, which will aid greatly on decision day.

When the time comes for decisions, however, we must realize that science and scientists, like their economic cousins, offer no special aid on issues of value. It is up to each of us to answer: Is it beautiful? Is it harmonious? Is it something that we want in our communal living room, the Earth? It is up to us to learn to say no when the time comes—to say that wildlife refuges are off limits to mining; to say that our last old-growth forests are too valuable for simple lumber and pulp; to say, if we choose—if we have the courage—that we desire to live in a place and a culture that takes seriously the idea of duties to future generations and to all forms of life.

8

Sustainable Life

And if, a century hence, our descendants have as much difficulty in comprehending resources or benefit-cost analyses or
environmental impact assessments as we have in understanding the significance of the philosopher's stone or the holy grail,
then and only then will the environmental crisis have ceased
to exist.

—Neil Evernden

The sense that I have, the sense that many people have, is
that the time has come to halt the continuing deterioration
of the Earth and, whenever and wherever we can, to stimulate processes of health and restoration. This is a conclusion
that a person can reach logically, by assessing the long-term
utility of humans. The Earth is our home and, to borrow
Marcus Aurelius's version of the age-old wisdom, "that
which is not good for the beehive cannot be good for the
bee." One can also reach this conclusion more emotively: a
declining ecosystem is ugly and simply seems wrong.

Land Health and Sustainability

In this chapter, I'd like to offer, as plainly as possible, several
ideas on how we might move to reverse this deterioration.

My belief, again, is that much of what needs to happen to stimulate land health must occur within us, in our heads and our hearts. We need to start thinking and feeling much differently about what we see around us and how we evaluate the things that are going on. Part of the work that's needed will involve further pruning of the old conqueror ideology. But part will involve embracing new standards of quality and beauty—and this can be a satisfying, energizing part of the course that lies ahead.

To respect the Earth's health and allow the Earth to restore itself, a definition of health is needed. Health could mean an absence of human influence, and it does mean this in the wilderness image. But when humans and their aspirations are added to the picture, as of course they must be, health is much more difficult to define. At one time ecologists thought that we could define land health in terms of stable, climax ecosystems. But we now know that nature is inherently more unstable than this—or at least its patterns shift and are far more difficult to discern—and no easy definition exists. A healthy land, it seems, is moving, shifting, and evolving.

If it is difficult to define health exactly—be it the Earth's health or our own—many of its traits are easy for us to spot. A physician checking our health will take our temperature, check our blood pressure, do a few lab tests, and the like, all to look for symptoms of good health. In much the same way, we can play land doctor and check some of the land's vital signs. A healthy land is one that is rich in species, both animals and plants. Over a period of not many years, species keep their populations in check, and they avoid degrading their food sources and nesting areas. Bare soil is rarely seen on healthy ground, and visible net soil erosion is limited to special natural settings, mostly arid ones. Healthy soil is rich in microorganisms and healthy water is usually drinkable, even when silt-laden and discolored. One useful indicator of health is for a land parcel to display within its bounds the full cycle of birth, growth, and decay. A healthy land is one in which vegetation and animals die naturally and, in their death, participate in new life.

The difficulty in defining land health with real exactitude stems in part from our limited understanding of nature's ways. We don't know all that nature does when it is healthy, and we therefore lack the knowledge (and the time) to spot all of the signs of ill health. To deal with this ignorance will require within us a good measure of humility and caution; it will require us to rely on educated instincts and to use values to help decide, conservatively and humbly, what is suitable for a place and desirable for a culture.

One way we can begin to regain harmony with the land is by recognizing the most obvious signs that the Earth's health is deteriorating. It is bad enough that we have rid the Earth of many species; it is worse that we continue to lose more of them. It is bad enough that many species have lost much or nearly all of their natural habitat; it is worse that their ranges in many cases are shrinking even further. Most of North America's wetlands are gone, and more disappear daily.

If we are to live in harmony with the Earth, if we are to stimulate land health, our goal must be to develop ways of living that are sustainable. If we can learn to live in ways that we can pursue indefinitely, the land around us will likely be healthy. Continued degradation of the Earth, plain and simple, is not sustainable, however it occurs. The continued loss of wetlands is not sustainable, nor is the loss of species and wildlife habitat and free-flowing rivers. In each of these cases and in many others like them, the Earth's stockpile is diminishing, and we can't for long keep cutting into the limited principal.

The best way to stimulate land health is to allow the Earth to cleanse itself. Our role as humans is not to fix the Earth but to fix ourselves, and the way for us to do so is by seizing on the goal of sustainable living. Let us examine all that we do and ask ourselves in each instance: Is this something that we and our descendants can continue forever?

Can we repeat this act, again and again, without harming the natural world? If the answer is no, if the act creates a new loss or aggravates an old one, then one day we must stop it. And the sooner we stop, the sooner we make the transition to lasting ways of life, the healthier we, and the Earth, are likely to be.

Aldo Leopold sought to capture this notion of sustainability in his land ethic. His ethic condemned as wrong anything that disrupted "the integrity, stability, and beauty of the biotic community." Leopold wrote in a day when unaltered nature was presumed to possess a degree of stability that we now think is absent. But his basic message remains true: we must end the proliferation of human-caused disruptions of nature. Broadly stated, our moral task in dealing with the planet is to expand upon, and find methods to implement, this simply stated injunction.

When we talk about ecosystem integrity, we are not talking about an ecosystem in which humans are absent. But neither are we talking about a definition of ecosystem in which all that humans do is considered natural and, hence, is viewed as a component of an ecosystem's integrity. Humans have made their marks, and the marks, or at least most of them, can stay. Integrity means an end to further marks, an end to further human encroachments into the domains of other species. It means an end to further conquests of nature.

Because so much of what our culture does is nonsustainable, the changes that need to come about will be many. Ignorance will always be a traveling companion, which means we'll need to make lots of calculated guesses. We'll need, in truth, to aim high when setting goals and to err on the side of caution by cutting back more than may really be needed. Perhaps we can succeed; it is clear that we must try.

Part of the education process that lies ahead will be to get people to identify resource-use practices that are obviously nonsustainable. Disrupting wilderness is one example easy to grasp. Building more dams on free-flowing rivers is another. These processes cannot be continued, again and again, because the original stock is only so large. Placing toxic wastes in the ground is a third example—at least when the wastes add to and do not replace the existing burden. Even ordinary community landfills are nonsustainable, at least when they are dry landfills designed to hold wastes over the long term, or when they contain compounds that one day will pollute groundwater, or simply when they contain materials that degrade in time frames of centuries and millennia. The age of plastics, after all, is still only a few decades old. Where will we be, garbagewise, five hundred years into this new age? Expanding cities, of course, also are not sustainable, and at some point soon we must draw the geographic line.

Whenever the issue of limits is raised, many people begin to squirm. For many decades our economy has been based on the continued consumption of more land and more resources. If we say no to more consumption of land and to more alteration of the raw, will not our lives and economy suddenly sink? The answer to this question is likely to be yes, but it is a conditional yes, an answer that we can alter if we choose to change the ways we calculate our gains and interact with the land.

Part of the forthcoming conversation will be about the idea of quality, what it should mean and how it will appear in our lives. The issue of quality, Wendell Berry tells us, "is always the same as the issue of propriety: how appropriate is the tool to the work, the work to the need, the need to other needs and the needs of others, and to the health of the household or community of all creatures." Quality is a place-specific, context-specific idea, particularly when applied to nature. As we decide what to do, and where to do it, human desire must become a far less important determinant. We must first figure out what things can be done in a given place consistent with the continued health of the surrounding natural system. To the extent we can make it so, nature must be our measure and model. Watching and mimicking nature will mean, among other changes, new rules of conduct: Do not build fish hatcheries; clean up the rivers. Do not scatter poisons to protect the flocks; add natural predators and guard dogs. When fire is part of the natural landscape, allow it to burn if possible, and get out of the way.

Nature as measure is an old idea, embraced over the years by many cultures and many successful land stewards. In modern Western culture, the idea is largely new, and to date we have brushed it away or at least not taken it seriously every time it has come up.

Explorer–geologist John Wesley Powell mentioned the need to listen to nature in the 1870s, after he gained a national audience from his popular exploits on the Colorado River System. Powell was an acute observer—acute enough to know that aridity was the West's most distinctive natural feature and that it should shape what humans did there. Lands west of the hundredth meridian were largely unsuited for crops and should be used for grazing, if at all. Water management and watershed protection are vital concerns in an arid land, and Powell proposed that political boundaries correspond with watersheds. Change political boundaries, he said; change land–use practices, change land laws, change our expectations; change these things and more, to bring them into line with what nature has provided. Make nature the measure.

All that Powell recommended made great sense, and all of it was rejected. Instead, the conqueror remained in charge, with all of his optimism, boasting, and insensitivity to the realities of nature. Political and land boundaries were set by the surveyor's azimuth, not by watershed lines. The plow broke the dry soil, and dust storms and ruined hopes followed close behind. Early, well-funded pioneers seized the prime water holes and, by controlling the water, soon dominated the lives of the ranchers and farmers who came later. To this day, rivers that cross state lines provoke jealous disputes, and, because the boundaries are artificial, each state is encouraged to divert and pollute before the next state does. So abused today is Powell's Colorado River that its water arrives at the Mexican border not just undrinkable but largely poisonous for crops. And still we are slow to learn. Instead of halting destructive irrigation practices and cleaning the Colorado River, we take the fish hatchery approach: we build expensive desalinization plants so that the water flow into Mexico is at least not fully unnatural. Desalinization requires massive amounts of energy, which means new power plants, which mean more air pollution.

Health, Value, and Beauty

One step that will help us greatly will be to develop new ideas of what we mean by healthy, valuable, and beautiful. These changes need to occur first in the private realm of our hearts and then in the public realm. As we begin to reestablish emotional ties with the land, we'll likely experience changes in the things that we sense are good and in the things that we want. If this happens, we might well find ourselves voluntarily and happily departing from many of our old destructive ways of living. Nature embraces diversity and polyculture. As we come to value what nature values, uniformity and monoculture may strike us as unsightly. Nature covers bare soil, and we too may see bare soil as wasteful and unhealthy. An uncovered farm field in winter may become a badge of ecological insensitivity.

We can illustrate this process by seeing how shifts might happen in a few particular settings—keeping in mind that these are only a few of many possibilities.

Several environmental problems stem from efforts to grow bluegrass or bermuda-grass lawns in the arid West. Bluegrass requires lots of water, which means irrigation, pumping, and expensive, energy-consumptive transport systems. Chemicals are needed because bluegrass is not well suited to this environment. Irrigation means runoff and return flow, which carry chemicals and the soil's natural salts into waterways, which become polluted. When we extract and use water like this, streams become less usable by wildlife, and aquifers decline. Over time, the chemicals harm the soil, and heavy irrigation brings natural salts to the surface that can weaken or even kill plants. Grass, of course, must be cut, and gasoline–powered lawnmowers are, pound for pound, highly productive sources of air pollution.

Why do we do all of this? We do it, evidently, because we think lawns are supposed to be grassy and because bluegrass strikes us as beautiful. Maybe bluegrass reminds us of the old place back East; maybe it's just custom and habit, or a long–established symbol of good living, or evidence of good breeding.

Imagine what might happen if our senses of beauty shifted. What if beauty came to mean something that is natural and well suited to a place; something that blends in with nature and helps establish a link between people and the ecosystems in which they live? What if, one day, a bluegrass lawn in the desert becomes another badge of the ecological bandit, something affirmatively distasteful because it is so obviously out of place? If bluegrass becomes ugly and desert plants gain new beauty, out bluegrass might go—along with the water, the chemicals, the irrigation lines, the salinity, the water pollution, the decreased wildlife habitat, and so on down the cascading route that environmental disruption typically creates.

Bluegrass is not the only plant that has little place in the desert. Grave environmental problems also result from our vast efforts to make the desert bloom with alfalfa, corn, cotton, and simple pasture grasses. For corn as for bluegrass, arid lands are harsh, deadly places, and corn grows only by continuous efforts to alter and, over time, to harm the fragile desert biome. It would be one thing if our nation were short of these crops and no better places were available to grow them; in that case, irrigated agriculture might make sense. But these crops are abundant, sometimes overabundant, and lands far more suitable to grow them stand idle.

The stimulus for much of this irrigation came from the U.S. Bureau of Reclamation, which sought through massive engineering and land alteration to turn dry lands into wet ones. In no sense, of course, was the bureau reclaiming lands; it was converting lands to nonsustainable uses to which the lands were and are ill suited in the long run. For the generations that come after us, the bureau's principal legacy is likely to be damaged soil, damaged rivers, and expensive reservoirs that are clogging with silt.

One of the reasons for our communal fondness for the blooming desert is that we derive from it a sense of a (seemingly) successful mastery of nature. Where there once was (we thought) nothing, there now are, through human efforts, rows and fields of productive crops. If there is any one thing we could do to improve our environmental plight, it is to eradicate this outdated desire for power and control. If there ever was a battle with nature, we've won it; we have developed weapons—our tools—that can reshape the Earth. Instead of taking pride in our ability to manipulate the land, we would be wiser to develop a sense of pride in our ability to show restraint. We need to take pride in our growing ability to foster the wild and the natural, and to bring as much of nature into our lives as possible.

The Earth also will benefit if we can rethink the indicators we use to gauge our human community's health. In personal life, we often know that quality and quantity are different issues. Imelda Marcos had too many shoes, and her greed and excess were signs of ill health. When it comes to public life, however, we often forget what we know. We assume too readily that more is better and that most means best. Our tendency is to rely heavily on economic indicators, with higher numbers equated with greater communal well-being. But higher and more are often ambiguous signals, signs of loss as well as of gain.

The statistic of new housing starts can serve as an example of the ambiguity in our usual signs of social health. The government regularly tells us about new housing starts and whether they are up, down, or level. The assumption seems to be that a downturn in starts is an indicator of ill health and that an increase is good. But more housing starts means, in most cases, more land devoted to housing and less land available for farms and nature. New houses will often mean more acres consumed for housing while old houses are abandoned, and the net effect of this on the Earth can be negative. Homelessness and undersatisfied housing needs— these are the problems that we face, and their existence is a clear sign of communal ill health. But more new houses may or may not address these needs; they may or may not make us healthier. Fewer housing starts may mean that people are caring for older houses and that communities are becoming more stable—and this, certainly, is good. We can celebrate declines in homelessness, to be sure. But all other things being equal, we can also celebrate reductions in housing starts, for fewer starts means a healthier Earth.

Whenever we hear reports of economic indicators, we need to be far more probing and inquisitive than we have been. The satisfaction of legitimate needs can be good. But higher production by itself is an ambiguous figure. It means earthly loss for certain and may or may not bring net human gain. Indeed, what we really should aim for, and celebrate when we achieve it, is the satisfaction of housing (and other) needs *without* consuming vast resources and, particularly, without exploiting more land and disrupting more ecosystems. Devoting vast acres to new urban sprawl should be an event worth mourning, even if some good has come from it, for the loss is great. It should be a source of embarrassment, a reason for finger-pointing and name-calling.

Newscasters can help emphasize this message by celebrating the ways in which less is clearly better. Less garbage to throw away is better. Lower energy consumption is better. Fewer acres needed for industry and agriculture—these, too, are better. If we want higher numbers to mean something good, we should be more sensitive in picking them. Since the early part of this century, the elk population in the lower forty-eight states has increased, through strenuous efforts of conservationists, from about one hundred thousand to over

five hundred thousand. Over these increases, we can legitimately rejoice.

As we contemplate our lives and our economy, we must
separate the issues of production and needs satisfaction, for
they are altogether different. What we need to be asking is:
Can I do as much with less? In many areas of life we need
lower inputs, lower resource consumption, rather than higher output. Agricultural researchers continue to push for ways
to produce more food, calling for more chemicals, more
machinery, and, today, even computers. Again, this push is
destructive for the Earth, if not also for farmers and farm
communities. Basic farm crops are all in oversupply, even
with subsidies that encourage farmers to reduce production.
As a community our need is for lower inputs with more or
less the same output, and success on this front is what we
should be celebrating. Let the push be for fewer chemicals,
less energy, less plowing, and fewer acres in production.

Throughout the economy we can pick examples of misguided goals and misdirected incentives. Our electric utilities,
to pick one example, are positively encouraged by regulators
in most states to produce and sell more power, when our
nation's need is to do precisely the opposite. Utility rates
typically still allow companies to earn based on their capital
investments and on the volume of their sales. To no surprise,
construction continues and power usage goes up. Again,
needs satisfaction and production are much different issues.
Our signs of land health should be conservation and lower
production; utility rate structures and other financial incentives should be designed accordingly.

In many ways, lower inputs will bring considerable sav-
ings, as the case of energy usage illustrates. For the typical
community, marginal expenditures on conservation are far
cheaper, watt per watt, than expenditures on new power
plants. In the case of water projects, the story is even more
clear. And these calculations would favor the conservation
option even more distinctly if they took full account of the

planetary exhaustion that higher consumption almost always brings.

As we reassess our practices a healthy practice must come to mean one that is gentle on the Earth. It must come to mean something that is sustainable and harmonious with nature in the long run. Over time we may develop the sense that a business that destroys nature without some compensating restoration is receiving an undesirable subsidy. A company that destroys the Earth, even some part that it "owns," acts unscrupulously, competing unfairly with similar companies that are more respectful of the land.

Very often we'll hear the claim that a destructive practice can't stop because jobs are at stake. But a job that yields destruction brings no net benefit to the community, and it therefore makes little sense. After all, we wouldn't continue a war so that soldiers could keep their jobs. We don't leave drug laws unenforced so that sellers of illegal drugs can remain employed. The Earth is a common resource, our communal home. Alterations of any type should be allowed only if they appear good for the community when we look as far down the road as we can see.

In examining present and prospective job–creating opportunities, it may aid our thinking to develop an index that measures the destruction caused per job created. In timber harvesting and extractive industries, increased mechanization has brought about higher and higher index numbers. Today's logger no longer says: I need a forty–acre woodlot. He says: I need forty acres to clear–cut today and forty more acres soon after that. Some jobs we simply cannot afford to keep.

Using Nature

Just as we shall need to develop new concepts of what is healthy, valuable, and beautiful, so we must aid our quest for health and harmony by rethinking the ways that elements of nature are put to use and should be put to use.

When the pioneers headed west into new territories, they saw lands that seemed unused, awaiting the human touch so that they could produce benefits. To the Bureau of Reclamation in the early twentieth century, water left in a stream was unused and wasted, just like a tree allowed to fall in the forest and rot. Today we hear the same argument from timber companies and, sadly, from the U.S. Forest Service, which certainly knows better: we must cut old-growth forests, we are told, because they are going to waste. Gifford Pinchot might well have nodded.

Perhaps the first axiom of an ecological understanding is that all components of nature—all resources, in effect—are always in use, every day. Water left in a stream fosters wildlife, feeds roots, and, these days, flushes pollution. The mature tree rotting away in the forest feeds insects, provides a home for birds, and in time recycles its nutrients to enrich the soil. When we cut and remove a tree, we haven't put it to use; we have simply changed its use. Once we understand this, new questions arise whenever we ponder the wisdom of altering the land: Is the proposed change wise? Is the new use better than the old one? Can we sustain the new use, generation after generation? Do we, in our ignorance, know enough to grasp the effects of the change? When our knowledge runs dry, what does our deep-seated intuition tell us is right?

Elements of nature are always engaged in their prime and most foundational function: preserving a sustainable, harmonious ecological order. By and large, the more elements of nature that we can leave alone to do this essential work, the healthier the Earth will be.

Once we've broadened our definition of use, it is easy to see that resources can serve many uses at the same time. Forests hold and enrich topsoil and provide wildlife habitat; they offer recreational opportunities, for the soul as well as the body; they moderate rainwater runoff, thereby protecting watersheds and mitigating downstream floods and droughts. Forests also supply timber for human use and can continue to do so in moderation. All lands can serve multiple uses, and the federal statutes governing federally owned lands largely mandate multiple-use management. True, the U.S. Forest Service and Bureau of Land Management have both had difficulties implementing the multiple-use idea. The Forest Service is accused (on powerful evidence) of excessively favoring timber interests; the BLM (on even stronger evidence) of favoring grazers and miners. But the multiple-use idea has been planted, and it needs to spread to private lands as well.

By broadening our understanding of how resources are used we can see even more merit in the economic image of externalities and land permanence. When a forest is nothing more than timber awaiting the saw, the external effects of timber harvesting can appear slight. When we compare it to a multiple-use forest, the externalities abound. The removal of trees disrupts wildlife patterns and, to varying degrees, exposes soil to erosion and drying. As the trees are removed, the soil is deprived of nutrients. If the cutting is severe, rain-water that once sank into the soil can now run off the surface, leading to erosion, polluted streams, possibly floods. Likewise, when a wetland seems unused, filling it apparently causes no harm. When the wetland instead seems busy filtering pollution and incubating fish, the filling process causes major disruptions, with the ripples spreading widely.

When a particular parcel of land is no longer producing timber, the loss to us might well be modest. Timber from across the globe could probably do just as well; perhaps plastic substitutes could fit the need. But when a particular forest is unable to fulfill its other, more important uses in promoting land health, substitutes are far less possible. It is largely to protect these other, nontimber values that we need to use our forests in sustainable ways.

Owning Nature

From new conceptions of land's multiple uses should flow new understandings of what it means to own things privately.

Property norms today reflect virtually no understanding of how one acre is naturally linked to the next and how conduct on one acre inevitably concerns all others. Our property rules focus on privacy, security, and zones of influence to the exclusion of natural communities, natural links, and land health. With an enhanced awareness of ecology, we soon spot the ties and the breaks, the dependencies and discontinuities, that surround us in our altered natural settings. The ownership of land ought to entail obligations as well as rights, and the main obligations must be to the Earth. Ownership of land must come to be a privilege, bringing with it the full duties of membership in a community of stewards. When nature is the measure, the Earth becomes a common pool, and ownership lines, like political boundaries, jump out at us as the artificial, human constructs that they are.

Property laws distract us from nature's cohesion, and they complicate the job of stimulating the Earth's healing processes. In dealing with this problem, our principal task will be to redesign property norms so that they are more limited and context specific. Landowners must face new limits on their rights to use land to the detriment of ecosystem stability. These limits can come in our laws, or they can come voluntarily, in our hearts and values, without resort to coercive rules. Most likely, some combination of these two approaches will be called for.

One issue will long nag our property system because it admits of no easy answer, and it is something we might as well face up to sooner rather than later. Our property norms must encourage owners to become place people, to exercise over their lands the kind of love and respect that characterizes a sustainable, long-term link, to be all that our rape-and-plunder conqueror was not. At the same time, the Earth's health on occasion will ask us to return a parcel or a resource to entirely natural uses. A parcel may be too important as wildlife habitat—as a breeding ground for migratory birds, as a feeding ground for a sensitive mammal, or as an integral piece of an ecosystem protection plan. Human uses of that parcel may need to shift to other, less sensitive spots.

When this happens, when for the Earth's health the owners need to move, the conflict can be painful. When land is inhabited by sensitive stewards, by place people, the loss can be considerable, and money is not the main issue. Almost by definition a steward becomes emotionally bound to the land in a way that has no ready dollar equivalent.

For some stewards, returning an acre to natural uses will be like giving up a loved one to God. The loss is painful, but the object of our love may be better off. As a servant of God gives when asked, so too a responsible citizen of the Earth must relinquish custody if the need arises. Compensation can cover market loss; only time can heal the rest. The community, the county, and the entire state will also be called upon at times to sacrifice, as land parcels here and there are removed from tax-yielding uses. We expect citizens to fight and die to preserve the nation's health. At times, equal sacrifices will be asked of us in our capacities as animals and ecosystem members.

Zero-Based Resource Allocation

Some years ago, zero-based budgeting became a popular notion in business and governmental circles. The idea was that each item in a budget should be examined, in full, each year—not to see how much it should be increased but to see whether it should be included at all, and at what level. In budgeting, as in other aspects of life, we tend to accept the present state of affairs as the starting point for thought. We assume that decisions made in the past have been reasonably correct and consider simply what minor changes should be made for the next season or cycle.

In dealing with the Earth, our tendency is to engage in the same kind of reasoning, and the effects have been hazardous. Often this issue arises in discussions about how to manage the nation's remaining natural areas, whether wetlands, wilderness areas, old–growth forests, aquifers, or other resources. In response to calls for preservation the prodevelopment forces raise the same hue and cry: it makes no sense to devote all these resources to preservation. Surely, they claim, half, or three–fourths, or nine–tenths of the resource should be devoted to human development.

As each year goes by more of the resource is lost, and the allegation remains the same—that we are locking up all we have. The defect in this thinking, plainly, is that if we continue to chip away, year after year, we end up with nothing left. This thinking ignores all that has happened in the past and focuses solely on the small amount of the resource that is left. Only by ignoring the past is it possible to talk about locking up entire resources.

In a zero–based allocation approach we would examine the countryside as it existed before human occupation. We would judge the amount that we should preserve, not by comparison with what little is left, but by comparison with what we once had: a land composed entirely of wilderness; a continent rich in wetlands, wild rivers, and ecosystems of all types; prairies that spread for thousands of square miles; wolf, elk, bear, and other large mammals roaming throughout the land. Over centuries of human occupation these resources have been consumed, altered, and disrupted, so much so in some instances that very little is left. A zero–based analysis would aim to get a better sense of balance, to come up with the best answer for what we should do with what remains.

Controversy still simmers in northeastern Minnesota over the restrictions on motorboats put into place by the creation of the Boundary Waters Canoe Area Wilderness. Motorboat proponents moan and groan that nearly the entire area is off limits to their high–power, high–tech recreation. The issue, as they present it, is whether the canoeists should get everything. In their continuing complaints, they call for multiple use and fair shares.

When we back up a bit, however, the Boundary Waters controversy looks far different. In Minnesota as a whole, probably ninety-five percent of the lakes or more are open to motorboats. Moose shy away from active motorboat areas, as do many nesting birds. Motorboats seriously disrupt the enjoyment and recreation of canoeists, which is why so many flock to the protected wilderness.

Once we wipe the slate clean we can ask, not why is so much off limits, but why is so little? How much of Minnesota should be open to motorboats and the disruptions that they inevitably bring? Should it be twenty percent? Fifty percent? Eighty percent? Without a great deal of reflection and eco-logical study, it is hard to pick a number. Given our igno-rance and the need to act cautiously, we might want to restrain ourselves as much as we reasonably can, picking a low number and sticking to it as much as possible.

We can depict this need for zero-based allocation with a simple model: Imagine an entire world composed of a single city block of ten lots. On the first day, all the lots are in nat-ural states, and the land is healthy. Humans show up and alter the first lot, tearing into it with the axe, the plow, and the backhoe. Over time the process of alteration continues, and the lots are disrupted, one by one. When the eighth lot is altered, questions are raised—should all the lots be dis-rupted? But two untouched lots are left, and it seems exces-sive to lock up all that remains. The ninth lot is turned over to the bulldozer, and only one is left.

This is, in fact, where we find ourselves today in the case of many parts of nature. We've plowed up or cemented over the first nine lots, and only the tenth remains. Sensitive observers say that the tenth must be preserved. But the forces of alteration shout "foul." We are locking up every-thing, they claim, which is a terribly one-sided kind of allo-cation scheme. Preserve a part of what we have, and use the rest. But, of course, when the rest has been used and only the small preserved part is left, the same issue arises, and the same proposal is made. Preserve a part of what we have, and use the rest. And so, step by step, our natural areas are lost. Our old-growth forests are lost. Our wetlands are lost. Fragile desert ecosystems and barrier islands are lost. Aquifers are drained. Mountain lion habitat disappears. And the story repeats itself, again and again, as we inch closer to the end of the block.

With zero-based allocation, we can draw a line and stick with it. This idea surfaced in the 1988 presidential campaign when then-Vice President Bush embraced it in his stance on wetlands preservation. Bush claimed to support a no-net-loss position on wetlands and agreed that an acre of wetlands could be developed only if an acre was created. The logic was simple: our nation has lost enough wetlands and could afford to lose no more.

Once President Bush took office, however, the no-net-loss pledge was soon postponed and subverted, and it is no wonder. No net loss is an idea utterly at odds with the orientation of nearly all modern political and economic thought. It is, in addition, a contagious idea, one that threatens to spread everywhere once it is allowed to enter the door. Wetlands are vital, but so are barrier islands, desert wilderness areas, and pristine northern lakes. Logically, no net loss is equally applicable to them all. The idea is sobering, for it says that we can't walk away from our messes and problems as has been the practice for generations. When the Earth's health is ignored, it is almost always cheaper in the short run to strike out and develop new ground than to restore and rehabilitate land already altered. And to politicians, developers, and many others, cheaper now means better.

As a community, we need to start saying "no more." It will soon be time to draw lines around communities and prohibit nearly all expansion outward. Future development should involve "in-filling"—making better use of space within the community's perimeter. Development outside the line might be allowed only on a no-net-loss basis; only if, for each acre lost to nature, another abused acre is restored to health or some other form of compensation *to nature* is made.

Until legal rules of this type are in place, those who alter natural areas might take on the moral duty of providing compensation. The person who builds a home in the open countryside might, as compensation, purchase several acres of eroded farmland and devote it entirely to natural uses. To facilitate the process of compensation, a land-for-nature fund might be created in each area or state, with each developer encouraged (and, later, obligated) to contribute money for restorative land purchases.

The idea of no net loss, of drawing lines and barring development, will doubtless provoke loud complaints about violations of property rights. Our long-held belief is that landowners have the right to develop their land, subject only to modest regulation. But this view of property ownership is one of the first ideas that needs to go. In a context-sensitive property scheme, development will be allowed only if and where it is consistent with a sustainable ecosystem, and compensation to nature may be needed to ensure that development causes no net ill effects.

Once zero-based thinking permeates our communal resource-use decisions, it may make sense for governments to undertake material buy-backs of private land to create large, healthy ecosystems. Serious students of the tall grass prairie tell us, ruefully, that no true prairie exists in this country. Scraps remain, here and there, sometimes over a hundred acres in size. But true tall grass prairie requires the participation of bison and wildfires and may well need an entire unaltered watershed. The Nature Conservancy has purchased a large ranch in the Osage Hills of northeastern Oklahoma and is attempting to piece together and restore enough land to recreate a true prairie. The idea is a splendid one; with governmental participation, it can and should spread.

Large-scale land restorations can catch the public imagination, the public's moral courage and pride, the way small projects do not. Imagine what it would be like to create a massive Buffalo Commons (to use one idea already raised), located in the dry grazing lands of eastern Colorado and Wyoming and western Kansas, Nebraska, and on into the

Dakotas—a grazing area like the first pioneers and Native Americans saw, expanding well past the horizon in all directions. A Buffalo Commons, say, of twenty or thirty million acres. Image a fifteen–million acre wilderness of coniferous forest, moose, wolf, and caribou in central and northern Maine. People could even enter these preserves and enjoy and use them in ways consistent with the dominant natural purposes.

Our nation enjoys thinking big and setting grand goals for itself. As the nation that invented the national park, it would be fitting for us to invent the idea of land reassembly for the Earth's sake. It would be fitting for us as a people to show the moral courage and environmental sensitivity needed to take grand, glorious steps to help the Earth reestablish its health.

9

Living at Home

If the way of Western civilization and Western religion was
once the way of election and differentiation from others and
from the earth, the way now is the way of intimate commu-
nion with the larger human community and with the universe
itself.

—Thomas Berry

In its dealings with the Earth, Western civilization is still in
its adolescence. It has grown up enough, in terms of its physical powers, to push and shove and to grab for what it wants.
But it is still learning what it means to woo the land and to
love it. As a species we aren't yet mature, for we haven't yet
settled into a way of living that is suitable and sustainable in
the long run. Like many adolescents we tend to exaggerate
our understanding. We are living off the Earth's principal, off
the assets that we have inherited, and we'll continue to do so
until we either exhaust the land or learn to find our way and
earn our keep.

   In the coming decades we need to become adults in our
dealings with the Earth. We need to face up to our limited
knowledge and admit that the older species—our parents, in
a sense—possess a great deal of inbred wisdom that we have

177

yet to learn. Mostly, we need to make ourselves at home in nature, but not as boisterous, uncontributing children and not as visitors who dirty the rooms and move on. We need to make ourselves at home as caring, mature adults who are settling in for life, for their lives and the lives of their children and grandchildren.

———

When I took over my land straddling the Hog Branch, I had a number of tasks to attend to. New jobs often call for new tools, and my first acquisition was a heavy barbed–wire cutter. A prior owner had grazed cattle in the woods and strung wire along the field edges, to keep the cattle from the corn. The local deer didn't seem to be slowed much by the extensive barbed wire, and I don't suppose the coyotes were overly concerned either. Still, the wire was out of place as well as hazardous, and I soon began my on–again, off–again project of cutting, recoiling, and recycling. The vegetative pattern here and there suggested to me that the grazing had ended some twenty years earlier, yet the barbed wire remained, slowly rusting. As landowners, we have been quicker to fence our fields than to unfence them.

Along the road on the property's south edge, a widening row of trees and shrubs encroaches on my flattest corn acres. The local extension agent asked: Didn't I want to rip out the vegetation to add another half acre to the field. No, I didn't, thanks; this half acre was already well used. My tenant farmer offered another idea: How about a pond on the property? I was tempted by this thought, and still am, but I want to consider the effects, long and hard. In this part of the country, ponds and lakes are typically shallow. The Hog Branch appears to carry a heavy sediment load, and I fear (among other things) that my pond would need frequent dredging. For now I'll enjoy the creek. If I procrastinate long enough, maybe local beavers will do the job for me.

I soon began what I intend to make a perpetual project: planting trees, berry patches, and varied perennial foods, for both people and animals. This place will be no English country estate with strict geometric plantings that contrast sharply with nature's apparent disorder. My fruit and nut trees are going here and there, in spots that seem well suited for growth, with deliberate efforts to mix the species. My hope is that disease and pests that hit one tree will have trouble finding the next one of the same species. This isn't to say, though, that my plantings are random. A fruit tree's first crop is the splendor of its blossoms, and I continue to seek out planting spots that will help me reap this early harvest.

My most important early task was to plant something on the twenty–three tilled acres that I was removing from grain production. I only had a couple of weeks to decide what to do, for planting season was at hand and spring rains were already taking a toll on the uncovered soil. The acres lay in four, irregularly shaped patches, two on the bottoms and two on the uneven high ground. I was daunted by the responsibility. My farmer was happy to plant whatever seed I gave him, so the how to do it was not the problem. It was the what, and the where.

My first thought was an ambitious one: start a prairie restoration project. A friend farmed land not far from my home, and on some of his acres he grew and harvested big bluestem and Indian grass. With the growing interest in prairie plants, there is a niche in the farm market for commercial growers of these native crops. My friend offered to provide these two seeds. But two species don't make a prairie, and I wanted to steer as clear as I could from monoculture. He warned me, from experience, that these two species are tough competitors. They are dominant species in the tall grass prairie; once well established, they resist intrusions that would add diversity. On balance this idea seemed unattractive, nor was I tempted by an offer from a prairie restoration company, which was willing to plant a far greater variety of prairie species for upwards of a thousand dollars per acre.

I had to think in other terms. Several phone calls led me

to a local representative of a helpful group, Pheasants Forever, which offers free seed to farmers and other landowners willing to devote land to wildlife habitat. I was in luck. I showed up for my seed, hat in hand, and was offered a mixture of three perennials—timothy, alfalfa, and a strain of clover. These seemed acceptable, but I probed, politely, about the prospect of more. Yes, it seemed, he did have another blend of three species, and yes, I could mix and plant them together. I was reluctant to press my luck—the seed, after all, was being *given* to me—but this was my chance to plant the acres fresh. Yes, he did have a partial bag of miscellaneous seed. He wasn't sure what was in it; mostly oats and brome, he thought. Yes, he could scrape out the bottom of his seed bin. By the time I left, I had a mixture of at least a dozen species. This was hardly a reproduction of nature's diversity. But it was time to plant, and this would have to do.

As it turned out, the seven-acre tract on the north side of the Hog Branch was no longer accessible to farm equipment, for the culvert had washed out with the spring run-off. My farmer offered to bring in his backhoe to rebuild a temporary bridge for this one-time planting effort. But I decided to decline. I'd take the lazy way; I'd let nature take charge of planting this field and watch the slow dance of natural progression. By late April, the field was already covered with annual "weeds," some knee high.

Finding Our Duties

Aldo Leopold crafted many of his ethical ideas during the 1930s and early 1940s, when the United States was suffering through the Depression and World War. This was a time when people began looking to the government for answers, particularly the federal government, and the New Deal's alphabet agencies and bureaus were a result. Given the times it was perhaps surprising that Leopold developed his land ethic as he did, that he didn't frame his critical essays as briefs for more governmental action. Instead, Leopold spoke to the reader as an individual. He challenged each reader to develop an ethical attitude toward the land. As he saw things, the solution to overgrazing, erosion, and declining wildlife was not a new governmental program, not more taxes and more governmental ownership.

The solution—or at least the part of the solution Leopold thought most important—was a new attitude toward the land within each person who controlled a portion of the Earth's resources. The basic problem, Leopold argued, "is to *induce the private landowner to conserve on his own land,* and no conceivable millions or billions for public land purchase can alter that fact." Leopold bemoaned the "clear tendency in American conservation to relegate to government all necessary jobs that private landowners fail to perform." Conservation, he believed, required a change of heart, a new orientation not based on economic self-interest. The issue was one of philosophy, not science or economics: "It is inconceivable to me that an ethical relation to land can exist without love, respect, and admiration for land, and a high regard for its value. By value, I of course mean something far broader than mere economic value; I mean value in the philosophical sense." In short, as Leopold would explain, conservation is not "something a nation buys" but "something a nation learns." And that learning had to take place one person at a time, as each individual developed an ethic of care. "The real substance of conservation lies not in the physical projects of government, but in the mental processes of citizens."

If the tendency in Leopold's day was to turn burdens over to government, that tendency is even more pronounced today. Government's role will still be large, as will the role of the free market. But both government and economic forces respond to what people want and value—if people demand throwaway convenience, the market will supply it, and government will consent. Individual values must change first, with more communal changes then following.

Our society is a consumptive, destructive one in which we each participate. We cannot point fingers at large businesses and government and claim that we have no part in their endeavors. Yet, if we each are guilty, it doesn't follow that we each must bear the whole responsibility. If we set out as individuals to save the Earth, we are bound to be disappointed. Saving the Earth is too big a task. A better goal, a more achievable one, is to minimize our own part in the destruction and in every way possible to make our ethical choices known. We would each do better to save our spirits by living individually as cleanly as we can.

As more individuals seek to live cleanly, there will be an increasing demand for more information, and information of a distinctly practical type. Our practices of daily living need to change so that we live lightly on the land. And to do this, in this technologically and economically complex age, requires a great deal of knowledge. One idea, developed at some length by environmentalist Lester Milbank, is to establish Environmental Extension Services, modeled after the highly successful Agricultural Extension Services. Agriculture agents hand out practical farming advice and respond to the peculiar questions and needs of individual farmers. Extension offices are typically affiliated with state universities, which puts them in touch with the most knowledgeable scientific researchers and even allows them to influence the agendas of future work.

Each community, if not each neighborhood, could use Environmental Extension Agents to help disseminate practical advice on everyday living. The best agents will be those most sensitive to the local ecosystem, those who have dozens and hundreds of tips on dealing with local pollution problems, local resource limits, and local wildlife needs. They will give out advice on solar energy, water conservation, and home building and rebuilding. A big area of work will involve recycling and the disposal of household and business items—and, earlier in the cycle, information on which products do and do not cause environmental harm. They will give out information on gardening and on methods of lawn and house care that require no toxic chemicals and costly energy sources. Perhaps model environmental homes can be built that demonstrate good practices in action. The information needs will be nearly endless.

For an Environmental Extension Service to succeed, we need to make the Earth's health a priority. Currently, our priorities are elsewhere. High agricultural productivity is a priority, even though for most crops oversupply is a problem. National defense is a priority, and we spend hundreds of billions of dollars per year on it. These priorities are easy to establish because we have no trouble grasping the ideas of food and famine, war and peace. The ideas of land health and sickness are harder to envision. Once we grasp what it means for the Earth to be healthy and make that health a priority, an Environmental Extension Service or something like it might soon arise.

The Spaceship and the Circle

In the mid-1960s, economist Kenneth Boulding popularized the idea (often attributed to Adlai Stevenson) that the Earth is like a large spaceship, hurtling through a dark, otherwise lifeless universe. Space exploration had caught the public imagination, and space capsules had begun sending back stunning pictures of the Earth taken from the moon. The spaceship image caught the imagination of many new environmentalists, and it probably did as much to stimulate action as any other image, before or since.

Boulding's spaceship image was able to capture and illustrate poignantly the shrinking size and increasing interdependence of all life on Earth. On a spaceship, human life is exceedingly fragile. All systems must operate properly or the astronauts are in grave danger. A spaceship has no place where garbage can be thrown and forgotten; all wastes are stored or recycled. If the air supply is polluted, there is no escaping it; if the water becomes polluted, the astronauts have no choice but to drink it. All is connected; all is dependent.

The spaceship image remains helpful precisely for its ability to capture this fragility and connectedness. It is helpful as well because it includes humans. Yet, over the past decade or so, it has been used far less frequently. Perhaps its declining popularity is due simply to overuse. It is too familiar, almost a platitude. Perhaps our few experiences with space capsule explosions have given the image a somber, pessimistic facade. But among ardent environmentalists the image may have lost currency for a different, more troubling, more valid reason.

Whatever its virtues, the spaceship image risks pushing us in the wrong direction. A spaceship is an artificial vessel, human-designed and human-constructed. It is a product of human cleverness and is, indeed, a symbol of scientific achievement. All that happens on a spaceship is logically predictable, and ignorance is nearly eliminated. On a spaceship, people are separated from all other life, indeed from the entire biosphere, and are going it alone. They have placed their ultimate faith in the cleverness of the scientific mind.

On Earth, where we really live, life is not like this. The Earth is far, far more complex than any spaceship; we didn't create it, and we don't understand it. On a spaceship humans count, and all else is a tool. On Earth, humans are one of many species, and if our animal welfare advocates have shown us anything, it is the thinness of the line that separates humans from other forms of life. In the spaceship what the astronauts do stays with them, within the cabin walls. On Earth our actions reverberate far and wide, in ways we cannot predict and cannot trace. In the spaceship, humans are in complete control. On Earth, they are not.

On a spaceship there is no recognition of human ignorance and no way to embrace it. Nature is withdrawn, taking with it its mysteries and mystiques, its beauties and its surprises. Somehow we need to bring the spaceship and space travelers back to Earth. We need to allow the astronauts to live and work in a home that is more natural, in a place that is part of nature and, thus, connected to all that nature is and means. We must allow the space travelers to regain their place in Gaia; their spot in the deep ecologists' horizontal chain of species.

The many images we have examined all offer helpful pieces for a composite image that fully captures the human predicament in nature. The images that we each select and construct should also help us see clearly the need for circular paths and circular motions in our lives. In the natural processes of the Earth there is rarely a beginning and an end. Things that go in one direction inevitably return to repeat the cycle. The rainwater moves out to the sea and, through evaporation, returns to the sky again to become rain. The soil nutrients flow into the plants, which die and decay to enrich the soil.

If we need an exceedingly simple image, we could hardly do better than to seize upon the unbroken circle. So much of what we do to the Earth is not circular—and we need to make it so. We extract minerals and materials from the ground and manufacture goods from them. But we typically fail to return things to the soil as we found them, so that the processes can begin again. We generate wastes in forms that don't break down, and we often dump the wastes in the ocean and in other inappropriate spots. To engage in sustainable living, we need to complete the product cycle. Like our products, our activities also must follow circular paths, bearing repetition, again and again. If an action is not one that we can repeat, if it is something that can only happen a limited number of times, then it is an action that consumes the principal, not the earnings.

When we develop a sustainable equilibrium with the Earth, the things that aren't sustainable should stick out by their foreignness. A provocative illustration of this foreignness was presented in a popular film made in the mid–1980s, *The Gods Must Be Crazy*. In the film, a Kalahari Desert clan finds a soda bottle that has seemingly fallen from the gods in the sky (and actually was, fittingly, pitched out as garbage by an overhead airplane). Unfamiliar with the object and finding its presence socially disruptive, the group members seek to get rid of the bottle. They bury it in the ground only to have animals dig it up. They throw it skyward only to have gravity bring it promptly back down. One Bushman is then entrusted to throw it off the edge of the Earth—to throw it away. As the bushman walks on and on to the Earth's edge, the humorous calamities begin.

Like the people of the Kalahari bush, we have trouble finding that illusive "away" where we can throw things and forget about them. Increasingly, "away" is in our backyards, and we don't like it. Not many decades ago the popular saying was "Dilution is the solution to pollution." But dilution didn't work, then or now, so we've had to keep looking for places to throw our bottles. There are still a few spots that seem like they are "away," but even these are getting closer and closer. The ocean has been away, but we are now confronted by fish with cancer and garbage on beaches. The deep underground still seems "away," and deep injection wells pump millions of gallons of toxic material into underground formations. But over time we will likely find some of this waste mingling with our groundwater, even percolating to the surface. We have reason to be afraid, whether the toxins are fifty feet below the surface or two thousand feet. We have already reached the point where no place on Earth is far enough away so that we feel safe using it for dumping highly radioactive waste.

This brings us back, as it should, to sustainability and the circular image. Dangerous wastes are the end of broken product and activity pathways. The answer is not to look harder, or deeper, or higher, for a mythical "away"—not to keep searching for the edge of the Earth—but to stop producing the wastes to begin with or to break them down into harmless component parts. What comes to us from the store should go back to the store, to be reused, recycled, or restored safely to the Earth.

—

Born and raised in central Illinois, I retain my predecessors' instinct for the edges. After pacing my land and gazing at the creek from all angles, I selected a spot for a cabin, to be built as my time and money allowed. The cabin site is on the hillside, just overlooking a sharp drop–off down to the Hog Branch where it joins a small tributary from the north and achieves its greatest width. Looking beneath the treed canopy from this site, I can see my weed–filled bottom field, which is, happily, surrounded and secluded by trees on all sides. This direction would be good for wildlife gazing.

The view in the opposite direction, toward the winter sunset, looks out over my highland meadow, which rises gently and irregularly into the trees in the distance. In this direction, too, I have much to watch. The summer wind ripples though the grasses. As the sun declines and the shadows lengthen, the deer wander out for twilight strolls, along with the occasional red fox.

The cabin will have broad windows facing both the creek and the meadow, windows that open to join inside space with outside. On one flank, a large screened porch, open in as many ways as possible, will be the space most used. A sleeping loft above will catch the evening breezes off the meadow to help push out the summer heat. A high–efficiency wood stove will yield what warmth the sun doesn't supply directly. I have spent many hours looking into composting toilets, natural ventilation, solar heating designs, solar electricity generation, and the like. Mine will be simplicity of the elegant kind.

My cabin will lie at the edge. Meadow and forest; dry land and creek; hillside oak and wetland willow; cultivated farm and untended field; human control and human restraint. It will be, too, for me and perhaps those who visit, on the cutting spiritual edge. My hope is that, in my dealings with the land, it can be on the forward edge ethically as well.

Becoming Place People

One good way to foster sustainable living is for each of us to become place people in our chosen places. As we each seek to find ways to live in equilibrium with the place that we call home, we will probably develop very special feelings for that place. We'll come to know it well and be familiar with the animals and plants that live there. We'll learn to distinguish signs of natural health from signs of illness, and we'll seek to do what we can to foster health. Place people are inclined to be the custodians of their special places and to look on them not in some cold, utilitarian, economic way but with greater appreciation for their many inherent values.

Place people also tend to embrace the notion of bioregionalism, which encompasses ways of governance and living that are based on biological and geological regions—just as John Wesley Powell proposed. They are more likely to make nature the measure of what should be done in a place and to recognize how interconnected their acres are with all that surrounds them. They are more likely to know that each place is unique.

Finally, place people are more likely to sign on to what Donald Worster has called the declaration of *inter*dependence, the manifesto announcing that every part is related to every other part in a complex web of vitality.

In disputes over landfill sitings and the like, we often see at work these days the so-called NIMBY syndrome—the increasingly predictable, increasingly loud response that greets any unpopular plan: not in my back yard. In a way, this response is selfish and irresponsible: we create wastes and pollution and then want to shove it out of sight. But at the same time it is a sign of health and caring, and we might just want it to spread. Those who take the NIMBY stance show concern over their acres. If they don't fully seek land health at least they seek to keep out the most powerful forces of deterioration. We would be better off in the long run—albeit inconvenienced in the near term—if in each neighborhood the NIMBY cry was strong and insistent.

Once the NIMBY chorus is successfully raised in all quarters, we'll be under strong pressure to change our ways. We simply cannot generate wastes that we are unwilling to put in our backyards. The answer to disposal is to avoid creation. We need to assume that whatever we use, create, destroy, consume, and waste will come from and return to our homeland. This cyclical understanding will be a central element of what might prove to be the best images we have of humans in nature—images that show people living, harmoniously, perpetually, at home.

To aid our transformation, it may help to foster the work of regional poets so that we may view the land through their sensitive eyes. As Justice Oliver Wendell Holmes said, writing of the prairie novels of Willa Cather, it is "a real gift to realize that any piece of the universe may be made poetical if seen by a poet." It is far easier to love and protect a place when we see in it beauty and poetry of the highest order.

Images of Home

We are most likely to care for the place that we feel is our home, and our strongest preservation instincts are those that surround the home. It is at home that we are most prone to think about the generations that came before and the generations that will come after. At home we are in our den, and we do not want to soil it.

Perhaps our best images, then, will be those in which we depict ourselves at home, in a natural home that is inhabited by us and visibly linked to other species and the forces of nature. We should put, not faceless strangers in the picture, but ourselves and the people who are close to us.

We can picture the entire Earth as our home, as indeed it is. But we may well be most inspired by a vivid image of a much smaller part of the globe. Let us be more particular, and

seize upon two places that might just offer the kind of inspirational images of a natural home that we fervently need.

We can begin with a vision of a large wilderness campsite, inhabited by several extended families for a lengthy,
even indefinite stay. We can set them in almost any wilderness area. But the more familiar we are with the wilderness,
the more vivid the image will be and the more power it may
contain. In my mind, the people are set in the canoe country
of northern Minnesota, the wilderness area that I know best.
The lake provides water to drink and fish for food. The forest
stretches to the edge of the campsite, bringing natural bal-
ances into the human community and offering fuel, protection, and more food. The people are at home, living in
harmony in a way that they can sustain.

Those who have camped in the wilderness and come to
love it can readily grasp the power in this image, whether
they leave the imaginary people in the canoe country or
move them to some other natural setting. In the wilderness,
we are close to nature, aware of our impacts, and sensitive to
all signs of disruption. The wilderness lake provides our
drinking water, and we must be careful not to pollute it. Fuel
comes from dead wood, and, as our campers follow the
wilderness ethic, they avoid gathering too much wood from
any single spot. As nature would have it, the forest is a mixture of species and ages. The wilderness campers do not
overfish, for the breeding stock must remain healthy. Their
refuse, too, must be handled well, to avoid attracting dangerous animals and contaminating soil and water. On a day-to-
day basis, all that goes on can be repeated again and again.

The power of this image comes in part because it taps
into the strength of the wilderness camping ethic. This ethic
admonishes us to leave our campsite as we found it, with no
lasting signs that did not predate our occupation. What we
carry in, we must consume or carry out. We only dispose of
readily degradable, natural materials, and disposal is done
safely, in the ground and away from the water. All that we do

must be circular, capable of repetition without leaving lasting marks.

For those who know and love a wilderness area, this image is also evocative because it offers a reminder of the fundamentals of our interactions with nature. When water must be carried up from the lake, our muscles make us acutely aware of every gallon we use, if not every quart and pint, and it is amazing how little we really need. When we do our work outside, we need very little artificial light. And while nature's temperature range can be extreme, we quickly learn that the comfort range is far broader than we often assume. In the wilderness, the sun is used to help warm us, and shade and breezes help to keep us cool. Natural forces are used to advantage, not defied, and the entire economy is strictly local. Needs are separated from desires, which are separated from extravagances. Nature sets the standards for what is beautiful and what is healthy. Nature is the measure.

Perhaps above all, it is in the wilderness that we are likely to feel closest to nature and to know that we are no longer contributing to the Earth's degradation. Having nature all around us, all day, in all ways, eventually makes us stop feeling like out–of–place visitors. We may need cycles of meals and rest, times of hot and cold, rain and shine, before the rhythms set in, but once the sense of equilibrium stands up within us, we can feel separate from the nationwide, worldwide assault on the Earth. With this feeling can come an otherwise unattainable soothing of the soul, a wistful, moral restfulness.

A wilderness stay is often followed by a period of reentry into "civilization," a period that can have quite similar effects in awakening us to the great differences in modes of living. In dozens of ways, the exiting wilderness traveler is struck by our rapid resource consumption and by how much we disregard the effects of our actions. Food comes from the grocery, we assume; water from the tap; heat from the furnace; and who knows or cares where the sewage and garbage go.

The challenge of the wilderness image is to find ways to carry into daily, civilized life both the wilderness ethic and the special sense of being part of the natural world. The wilderness image, of course, is artificial. But its value is not that it offers us a way of life that we can imitate, detail by detail. Instead, it focuses our attention on our fundamental ties with the planet, on the things that really count. The wilderness image distills these elements into simple, easily carried form.

The wilderness image doesn't tell us to cast off the tools and toys of the modern age. It says: Remember the wilderness campsite in daily life. Remember the differences between needs and whims. Remember how sensitive nature is and how careful we must be to keep it clean and vibrant. Remember the wilderness goal of leaving no lasting signs and living so lightly that those who come later will spot no degradation. Remember what it is like to know where all of our things come from and where, once used, they will go. Remember what it is like to work with nature, to use what nature offers and in the quantities that nature can afford.

Remember all of this, and translate it into modern, non-wilderness life. In every way possible, seek to gain, and to deserve, the moral solace that comes from a style of living that respects the Earth's integrity. In the modern industrial economy, this goal is, for any individual, a high one. But it is the right goal, the right guiding light.

Setting to one side this wilderness campsite image, we can turn to a second image with similar messages that may offer, for some people, greater energy and clearer vision. Let us consider this time a rural, pastoral scene, a family farm in a broken countryside of fields and forests, hills and plains. The farm is inhabited by a family of several generations. There is a large garden, fruit and nut trees, livestock, varied crops, ponds, woodlots, and pasturage. The farm is the family's home, has been and will be, and rightly used it will meet their basic needs for generations and generations. As observers, we notice distinctly the dominance of circular pathways on this farm, in all that goes on. The sun and the seasons set the tempos of life. All that is done can be redone, and done yet again, for the family uses natural fertilizers and reduces insect pests through crop rotation and intercropping. The land is kept covered, and well-tended livestock provide much of the farm energy. Heat comes from the sun and the woodlot; cooling from the shade, from the breezes, and from sensitively designed and situated living structures. Local knowledge is used in abundance.

This farm is "owned" only in relation to the outside world. Among those who live here and those who will live here the idea of ownership has little meaning. Each generation has the privilege of using and living on the land and the duty to pass on as good and as much as it got. Perhaps the farmer is a hunter and brings to the table the occasional duck and deer. The family knows where its food and water comes from and where its wastes will go. Steep slopes are never touched with the plow. Woodlots and field edges offer home and food for the wildlife that share the scene; wildlife furnishes a reminder of the horizontal kinship of all animals.

Like the wilderness campsite image, the family farm image offers senses of harmony, of balance, of sustainability, and of humans at one with the natural world. For some people, the farm image is likely to offer an alluring model of life, something to be imitated exactly. But most people are unlikely to find farm rigors to their liking, particularly the 365-day work schedule that livestock brings. For most people, the family farm, like the wilderness campsite, is an image that offers more general guidance, and we should use it for that purpose.

The farmer in our pastoral scene is, preeminently, a place person, and all that goes on is geared to the intimately known capabilities and temperament of the land. This is a particular farm, not just any farm, and what is done is what nature allows to be done. What comes from the land returns to the land. Productivity is geared to the needs of the family, and the land is asked to do little more. Future generations play an important role, and in a particular, expected way. The farm in this image may be unlike the place where we live, but the principles of living are the same. As on the farm, our interactions with nature must be capable of repetition. They must be sustainable. Nature, as always, must be the measure.

Like the wilderness campsite, this pastoral scene does not reject high technology and specialization—it provides the foundation on which they occur. The farm scene need not be an arcadian idyll, some bucolic rejection of the modern. The farm house might boast a high-tech solar heating system, even the latest satellite dish or a cranked-up stereo system. Its inhabitants might display all the shifting fashions of dress and personal appearance, of music and literature, of gender roles and intellectual discourse.

The aim of the pastoral image in an ecological age is not to bring change to a halt, much less to drive out all machines. The aim is to promote health—the health of the Earth and, through it, the health of humans. Like the wilderness campsite image, the pastoral scene draws attention to the fundamentals: clean air, clean water, protected soil, vibrant biodiversity, renewable resource cycles, respect for the unknown—above all, nature as the measure of what is healthy, beautiful, and wise.

Given this ecological focus, it should be plain how this contemporary pastoral vision differs from the eighteenth- and nineteenth-century pastoral images that served to help reconcile the established agrarian culture with the rapidly expanding technological mentality—the machine that brought smoke and noise to the garden. The new pastoral vision is aimed at achieving ecological equilibrium, not a social or economic utopia. It doesn't imply a rejection of all machines but a preference for certain machines over others; it favors the gentle machine over the environmentally destructive one, the solar panel over the coal furnace, the electronic communication network that undercuts the need for consumptive air travel. Within this framework of ecological health, science and progress can flourish; in fact, they'll be needed to help bring it all about. But progress needs to be of a qualitative kind, measured in efficiency, health, and well-being, not in terms of increased consumption and depletion.

This pastoral image can help channel future growth and progress by offering a set of guiding moral values. In important ways, ecological values resemble the norms that govern and elevate our dealings with one another. The social arena is dynamic and exciting, yet it contains firm bans on physical abuse, duress, fraud, and discrimination. These social norms, and others like them, provide moral anchors to help contain destructive, antisocial urges. We need the same kind of moral anchors in our dealings with the land.

For most people, as with the wilderness campsite image, the pastoral image is less a model for daily living—although parts of it can apply—than it is a way of evaluating communal practices and setting communal goals. In their interactions with nature, the town and the city are simply farmsteads writ large. Water, food, fuel, and other resources all come from the land, somewhere, and the return flow, clean or dirty, goes back to the land. The urban community must learn to live in a way that is as clean, as wholesome, and as sustainable as life on the family farm. The fundamentals, as always, are the same—water and waste, soil and air, wildlife and room for mistakes.

In the urban setting, to be sure, the individual cannot take full charge. Power is diffuse and technical issues are complex. Yet it is in the face of such complexity that simple images of right living can do the most good. If the details are complex in application, the goals are not. If the engines of government and business are necessarily intricate, the jobs set for them need not be. Let us raise up our simple images and point to them, saying, "This is how we want to live, using the fruits of nature and not the principal, allowing plenty of room for our planetary cotenants, avoiding bets that we cannot afford to lose, leaving the planet when we each depart a place that is as healthy and productive as when we arrived."

The wilderness campsite and family farm images offer us glimpses of people living at home in nature successfully and continuously. They are simple images of right living, and they offer us lessons and guidance to carry in our minds—wisdom to use and wisdom to share. They can offer us strength and energy for our hearts, enough, again, to use and to share. They are, to be sure, not the only images of this type, and the best image, in the end, may be the one that each of us develops individually.

But whatever image we might each select, let us keep it clear and put it to use. Let it shed light on our lives and infuse our values and our public discourse. Let it help us keep track of right and wrong and help us push ourselves, privately and publicly, in the direction of the right.

Let us all live in our natural homes, with what nature offers us and with the fruits and products of what we create, and let us do so through the generations and generations, again and then again.

Notes

Opening quotes. The higher processes. . . Willa Cather, "The Novel Demeuble," in *Willa Cather on Writing* (Lincoln: University of Nebraska Press, 1988), p. 40. *The Great Way is. . .* Lao Tzu, *Tao Te Ching,* trans. Victor H. Mair (New York: Bantam Books, 1990), 16.4–5.

page xix. State a moral case to a ploughman. . . Letter from Thomas Jefferson to Peter Carr, August 10, 1787, 12 *The Papers of Thomas Jefferson,* pp. 14–15. ed. J. Boyd (Charlottesville: University of Virginia Press, 1955).

page 1. Unbidden, unplanted, the Old Prairie. . . Dan Guillory, "Lost Gardens," in *Benchmark: Anthology of Contemporary Illinois Poetry,* ed. James McGowan and Lynn DeVore (Urbana: Stormline Press, 1988), p. 142.

page 8. We all gather the pearls. . . Quoted in James Woodress, *Willa Cather: A Literary Life* (Lincoln: University of Nebraska Press, 1987), p. 110.

page 9. Much of the damage inflicted on land. . . Aldo Leopold, *A Sand County Almanac With Essays on Conservation from Round River* p. 197.

page 10. one of the penalties of an ecological education. . . Aldo Leopold, *A Sand County Almanac With Essays on Conservation from Round River* (New York, Ballantine Books, 1970), p. 197.

page 16. a seedbed in the heart. . . Stephanie Mills, *Whatever Happened to Ecology?* (San Francisco: Sierra Club Books, 1989), p. 5.

page 16. There must be a mystique of the rain... Thomas Berry, *The Dream of the Earth* (San Francisco: Sierra Club Books, 1988), p. 33. *To change ideas about what land is for...* Aldo Leopold, "The State of the Profession," in *The River of the Mother of God and Other Essays*, ed. Susan L. Flader and J. Baird Callicott (Madison: University of Wisconsin Press, 1991), p. 280.

page 19. Man's power over Nature... C. S. Lewis, *That Hideous Strength* (New York: Macmillan, 1946), p. 178.

page 45. Property belongs to a family... Wendell Berry, "Whose Head is the Farmer Using? Whose Head is Using the Farmer?" in *Meeting the Expectations of the Land*, ed. Wes Jackson, Wendell Berry, and Bruce Colman (San Francisco: North Point Press, 1984), p. 30.

page 65. I suspect there are two categories... Quoted in Curt Meine, *Aldo Leopold: His Life and Work* (Madison: University of Wisconsin Press, 1988), p. 361.

page 93. Possibly, in our intuitive perceptions... Aldo Leopold, "Some Fundamentals of Conservation in the Southwest," in *The River of the Mother of God and Others Essays*, ed. Susan L. Flader and J. Baird Callicott (Madison: University of Wisconsin Press, 1991), p. 95.

page 95. retaining its primeval character... Wilderness Act of 1964, United States Code., Title 16, Section 1131.

page 100. We simply need that... Wallace Stegner, "Wilderness Letter," in *The Sound of Mountain Water* (New York: Dutton, 1980), p. 153.

page 105. Gaia theory forces a planetary... James Lovelock, *The Ages of Gaia* (New York: Norton, 1988), p. xvi.

page 113. If we want to know... Wendell Berry, *Standing by Words* (San Francisco: North Point Press, 1983), p. 66.

page 116. My aim is to show... Quoted in Brian Tokar, *The Green Alternative* (San Pedro: R. and E. Miles, 1987), p. 21.

page 120. As we acquire more knowledge... Albert Schweitzer, quoted in *Harpers*, July 1990, p. 33.

page 131. To keep every cog and wheel... Aldo Leopold, *A Sand County Almanac With Essays on Conservation from Round River* (New York: Ballantine Books, 1970), p. 190.

page 133. God guard me from those thoughts... Quoted in Wendell Berry, *Home Economics* (San Francisco: North Point Press, 1987), p. 116.

page 150. The hanging involved no question... Aldo Leopold, *A Sand County Almanac and Sketches Here and There* (New York: Oxford University Press, 1949), p. 201. *To include soils...* Aldo Leopold, *A Sand County Almanac and Sketches Here and There*, p. 204. *We abuse land...* Aldo Leopold, *A Sand County Almanac and Sketches Here and There*, p. 204.

page 151. A thing is right... Aldo Leopold, *A Sand County Almanac and Sketches Here and There*, pp. 224–25.

page 151. changes the role of Homo sapiens... Aldo Leopold, *A Sand County Almanac and Sketches Here and There*, p. 204.

page 155. And if, a century hence... Neil Evernden, *The Natural Alien* (Toronto: University of Toronto Press, 1985), p. 144. *that which is not good...* Quoted in Wallace Stegner, *The Spectator Bird* (Lincoln: University of Nebraska Press, 1976), p. 96.

page 158. the integrity, stability, and beauty... Aldo Leopold, *A Sand County Almanac and Sketches Here and There* (New York: Oxford University Press, 1949), pp. 224–25.

page 159. is always the same as the issue... Wendell Berry, *Standing By Words* (San Francisco: North Point Press, 1983), p. 51.

page 177. If the way of Western civilization... Thomas Berry, *The Dream of the Earth* (San Francisco: Sierra Club Books, 1988), pp. 136–37.

page 181. is to induce the private landowner... Quoted in Curt Meine, *Aldo Leopold: His Life and Work* (Madison: University of Wisconsin Press, 1988), p. 321. *clear tendency in American conservation...* Aldo Leopold, *A Sand County Almanac and Sketches Here and There* (New York: Oxford University Press, 1949), p. 213. *It is inconceivable to me...* Aldo Leopold, *A Sand County Almanac and Sketches Here and There*, p. 223. *something a nation buys...* Quoted in Curt Meine, *Aldo Leopold: His Life and Work*, p. 370. *The real substance of conservation...* Quoted in Curt Meine, *Aldo Leopold: His Life and Work* (Madison: University of Wisconsin Press, 1988), p. 402.

page 189. a real gift to realize that... Quoted in James Woodress, *Willa Cather: A Literary Life* (Lincoln: University of Nebraska Press, 1987), p. 302.

Index

GE
140
F74
1993